TOWARD A NATURAL FOREST

Kathy Sutter

TOWARD A NATURAL FOREST

The Forest Service in Transition

a memoir

Jim Furnish

Oregon State University Press
Corvallis

Illustration by Kathy Lutter
Map calligraphy by Caren Milman

The paper in this book meets the guidelines for permanence and durability of the Committee on Production Guidelines for Book Longevity of the Council on Library Resources and the minimum requirements of the American National Standard for Permanence of Paper for Printed Library Materials Z39.48-1984.

Library of Congress Cataloging-in-Publication Data
Furnish, Jim.
  Toward a natural forest : the Forest Service in transition : a memoir / Jim Furnish.
      pages cm
  ISBN 978-0-87071-813-7 (original trade pbk. : alk. paper) — ISBN 978-0-87071-814-4 (e-book)
1. Furnish, Jim. 2. United States. Forest Service—Officials and employees—Biography. 3. United States. Forest Service—History—20th century. 4. Forest reserves—United States. 5. Forest ecology—United States. 6. Forest conservation—United States. I. Title.
  SD129.F87F87 2015
  333.75′11092—dc23
  [B]
                              2014046158

Oregon State University Press
121 The Valley Library
Corvallis OR 97331-4501
541-737-3166 • fax 541-737-3170
www.osupress.oregonstate.edu

for
Bill Ford
a wonderful and true friend

Orono ME
1968-1971

Wash. DC
1989-1991
1999-2002

Ames IA
1968

Black Hills NF
1971-1976

Grand County
1976-1977

Bighorn NF
1977-1984

San Juan NF
1984-1989

Siuslaw NF
1991-1999

Contents

USDA Forest Service
Organizational Structure

Secretary of Agriculture

Under Secretary for Natural Resources and Environment

Chief Forester

Deputy Chief National Forests

Regional Forester

Forest Supervisor

District Ranger

Foreword

People make institutions. They create them, organize their goals and objectives, bond with like-minded colleagues within them, and can inhabit these social spaces to such an extent that they can become one with them. We are where we work.

But what happens when we begin to perceive flaws in the organizational mission, sense that we are uncomfortable with its prevailing beliefs, or discover we are at odds, in this sense, with our peers and even ourselves?

These questions—and the complicated answers they generate—are woven throughout Jim Furnish's superb memoir, Toward a Natural Forest. He has written a powerful, passionate, and engaging book about his lengthy and distinguished career in the US Forest Service. In truth, this text tells us as much about his evolving perspectives as it does about the federal agency and the shifting landscape, natural and human, in which it has operated since its founding in 1905.

Furnish makes the most of these transitions. And with reason, for his tenure in the agency—he started as a seasonal employee in 1965, and left the Forest Service as one of its deputy chiefs in 2002—spans arguably the most raucous, tumultuous time in its history. Initially a low-level worker with little clout, Furnish in time and with success would become an active agent of change with considerable authority. From these different positions of power he bore witness to the degree to which agency culture was dominated by a "Timber is King" mindset, a dominance that at first made sense to him but over time began to diminish in its hold. By deftly narrating how he slowly, painfully began to question his employer's ethos, by articulating why he felt compelled to voice his concerns about its raison d'être, Jim Furnish helps us recognize the deeply personal nature of public service.

Born in 1945 in Tyler, Texas, Furnish came of age trailing along with his geologist father who each summer led field trips to the Black Hills in South Dakota and Wyoming's Bighorn Mountains, students and son in tow. "From about the age of eight, and over the course of several

1

succeeding summers, I pretty much did whatever he did," Furnish recalls. "He strived to teach me what he knew about hunting, fishing, canoeing, birding, trees, geology, and many other things. Some of it stuck, and all of it affected my heart and my attitude toward our Earth."

An avid observer, Furnish learned as well to love the West—its jagged topography, big sky, star-studded nights, and the redolent scent of pine. Forestry seemed to offer him a way into this evocative landscape, a profession that provided a fine livelihood and a better life. So, as he studied for his degree at Iowa State University, he spent summers on the Umpqua and Malheur National Forests in Oregon, field training that led to his first full-time position with the Forest Service as a researcher in Orono, Maine. Down East was not Far West, but Furnish gained insight into the power of the wood-products industry in the Pine Tree State, a lesson that resurfaced later in his career as he confronted the agency's attempt to mimic industry's profit-driven maximization of timber resources by clear-cutting the national forests.

He did not immediately realize that a ramped-up policy of aggressive harvests had deleterious consequences. One of the virtues of this book, in fact, is that Furnish does allow his readers to watch his evolution—fitful and partial—as it unfolded in real time. Early in his career, for example, he saw the ecological impact of extensive, heavy logging on the Black Hills and Bighorn Mountains, but did not instantly appreciate its significance. One turning point came after trying to regenerate a logged-out lodgepole pine forest high up in the Bighorns. Replanting efforts failed repeatedly, suggesting that the political demand to get out the cut had run into a serious ecological restraint. Not all woods could or should be reduced to board feet.

These insights were not Furnish's alone. He rightly credits outside forces challenging the Forest Service's convictions with shaping his emerging awareness of the decided limitations of the agency's (and his) actions. "By the late 1960s," Furnish observes, "the Forest Service was engaged in a battle for the soul of public lands, and a decades-long, slow-motion collision with a robust and rising environmental movement." The shock waves were felt in Congress, which beginning with the passage of the Wilderness Act (1964) and continuing with the 1970s Clean Air and Water legislation, the National Environmental Policy Act

(1970), the Endangered Species Act (1973), and the National Forest Management Act (1976), placed greater restraints on the agency's land-management schemes. They were manifest too in combative public meetings over forest plans, street demonstrations that opposed logging operations, and countless lawsuits filed to protect spotted owls, marbled murrelets, and salmon.

Like some of his peers, Furnish reflected on the public clamor and started to reconsider whether timber should remain the driving engine of the Forest Service's mission. These dissenters' ranks were bolstered by new employees trained in ecology, hydrology, and wild-life management—all essential skills if the agency was to comply with the new federal environmental regulations—who also raised difficult questions about once-uncritically accepted tenets of in-house cul-ture. Dubbed "combat-ologists" for the bruising internal wars they engaged in, these individuals were among the first members of the Association of Forest Service Employees for Environmental Ethics; by its name and through its members' whistle-blowing activism, FSEEE (the words *association of* were later dropped from its name) was calling out the Old Guard.

Yet what it meant to be in the New Guard was not clear. For Furnish, the most significant opportunity to clarify this situation, and to test the principles of ecosystem management on the ground, occurred between 1992 and 1999 while serving as forest supervisor of the Siuslaw National Forest. Located along the central Oregon coast, the Siuslaw is a tremen-dously productive temperate rainforest. Its managers had been getting out the cut for decades, ringing up substantial receipts for the Forest Service and racking up sizeable environmental deficits. Furnish care-fully narrates the personnel roadblocks he had to clear, the new hires that came on board, and the local labyrinths he had to navigate before he and his reconstituted staff could reduce harvesting levels, close down logging roads, restore battered riparian habitat and once-fertile estuar-ies, and protect endangered species, building a management framework whose principal goal was the regeneration of the natural forest.

Success on the Siuslaw propelled Furnish to Washington, DC, where he served under Chief Mike Dombeck as the Deputy Chief of the National Forest system. It was a win-lose proposition: Furnish's promotion was one of many signals that the thirty-thousand-person agency was trying to change, top to bottom; but it also generated an intense internal struggle of wills that was mirrored in pro-and-con pressures from Capitol Hill, the courts, and the body politic. The timing was also tight: Furnish arrived in the last two years of the Clinton administration, so any of the reforms Dombeck hoped to enact, including the far-reaching executive ruling to permanently protect roadless areas within the national forests, had to be completed before a new president was inaugurated in January 2001. Furnish and his staff beat that deadline on the roadless rule. Although the incoming President, George W. Bush, immediately set it aside, in 2012, after a decade of legal wrangling, the Tenth Circuit Court of Appeals countermanded Bush's action, and the Supreme Court let the lower court's decision stand. "Reading their opinion," Furnish confessed, "was sweet vindication."

This validation did not mean that the struggle to implement an ecosystemic ethos as the paramount principle defining national forest management had been upheld. With Dombeck's resignation in 2001, Furnish found himself an odd man out and shortly thereafter left the agency. Offstage, he continues to argue for the central ideas that over time had come to define his professional engagements. Indeed, the final chapter of *Toward a Natural Forest* offers a well-defined path forward for the Forest Service that would start with it adopting a sustainable land ethic to shape its actions. Furnish's vision rejects rampant exploitation and extreme preservationism in favor of what he describes as a middle ground, "where fundamental stewardship humbly, modestly, and sustainably utilizes what is produced and provided by nature. These beneficial and essential environmental services can be enjoyed in perpetuity as long we do not irreparably damage our forests, soils, water, and air."

That's a tall order, made even more daunting in a climate-distorted era, but that's also why Furnish is right to call the question, and right to urge those inside and outside the Forest Service to keep pushing the agency to live up to its credo: Caring for the Land and Serving People.

Char Miller

Preface

I chose to spend our US Bicentennial, July 4, 1976, backpacking Colorado's Never Summer Mountains, which form Rocky Mountain National Park's western border. Remnant ice at midsummer on Lake of the Clouds was a reminder of the long winters and brief summers in the high peaks. The lake's cobalt waters slowly emerged from a long slumber, with an ice shelf hugging the western shore. As I ambled through the alpine high country above timberline, enjoying the freedom of being small in big country, a faint tinkling sound caught my ear. I approached the margin where open water and ice met and discovered the source of the sound. Small waves from an afternoon breeze roughening the water lapped against the edge of the ice, gently separating long daggers of ice that danced and bumped in the water.

Each icicle a chime now, they numbered in the thousands. From the waters emanated a sublime symphony of tinkling bells, delicate and magnificent. Exquisite.

Winter lost its grip, and the ice did, too, breaking up as it slowly disappeared. But the loss of ice was accompanied by the gain of something beautiful. With nature, this miracle of death and rebirth happens every year, part of nature's cycle, comforting in its regularity. Human endeavors also confront change but are often accompanied with grief and stress instead of comfort. So it was with the breakup of the US Forest Service's old order.

The Forest Service of the 1950s was heavily populated with men of righteous zeal, winners of a great war, the kind described by Tom Brokaw in *The Greatest Generation*.[1] They aimed to log national forests aggressively for a wood-hungry nation. By the late 1960s, the Forest Service was engaged in a battle for the soul of public lands and a decades-long, slow-motion collision with a robust and rising environmental movement.

How did the Forest Service in which I spent my career confront the new while releasing the old? Even though I believe the old Forest Service and many of its cherished traditions perished, it heartens me to

5

hope that a new Forest Service might awaken to make music with the icy shards of its past.

I want to tell the story of this breakup from the inside out through two intertwined tales. The first tale involves my beloved Forest Service, which, stewarding a natural world with the best of intentions, managed wildness unto submission and, perhaps, death. The second tale involves my personal transformation as a forester, in my guts and in my blood. I began my career accepting without reservation the prevailing ethics of the Forest Service, then began to question, confront, and change them, and finally arrived at a place where I felt I was unwelcome and had to leave. I tell these tales through experiences involving smaller issues that exemplify larger battles over the social and environmental values that shape resource management and policy. I conclude with my take on the message I heard in the symphony of landscapes that speaks to us all.

My Forest Service career began in 1965 in Tiller, Oregon, and ended in 2002 in Washington, DC. I dreamed of working for the Forest Service ever since my older brother returned from a 1960 summer in Quinault, on Washington State's Olympic Peninsula, full of great tales of legendary forests and foresters.

In 1965, the Forest Service was highly regarded and highly motivated, doing big things in national forests. Trees were cut and moved into the mills and homes of America at unprecedented levels. Roads penetrated vast reaches of forest for the first time.

By 2002, the Forest Service had been hard at work for nearly one hundred years, but it entered the new century a more humble agency, distrusted and even vilified by an emergent environmental constituency that saw little evidence that public lands were being managed honorably. For more traditional allies and many agency employees, the Forest Service just seemed to have lost its way. The once-pleasant atmosphere in almost every Forest Service office, a product of "doing God's good work," had been replaced by a nervous wariness. We no longer felt appreciated, comfortable, or secure.

The Forest Service had lost its grip and was breaking up. Could something beautiful be gained from this loss? By the end of my career, I could answer yes, but I believe I am unusual. I remain troubled that for most people who work in and with the Forest Service, the answer remains no.

With courage and wisdom exhibited for more than a century, America has ambitiously set aside a vast complex of public lands—fully one-third of our country—for national parks, forests, wildlife refuges, and other public purposes. This penchant for conservation contrasted with much of the world's restive exploitation of natural resources. Now many other nations have belatedly embraced this concept. We see parks and reservations blossoming all over the globe. These lands have become increasingly important in the struggle to address an exploding population and mitigate the ensuing exploitation of resources. Management of our national forests mirrors, in microcosm, a much larger issue: how best to balance the needs of humans while adequately protecting the inherent integrity of the environment?

I worked in forests and mountains from coast to coast, and, as an employee, migrated from the bottom of the Forest Service to its highest ranks. It seems reasonable that the Forest Service—charged with managing spectacular mountains and forests comprising fully 8 percent of all lands in the United States—would hunger to lead the society it served in addressing important environmental issues. Herbert Kaufman extolled the Forest Service in *The Forest Ranger*[2] as an effective, resolute agency fully committed to delivering on the hopes of an expectant public. But forty years later, at Yale University, the womb of American forestry, I heard Kaufman say that he had also cautioned that the Forest Service's monolithic character might prevent it from anticipating and adjusting to change.

Sadly, as events unfolded in the latter twentieth century, the Forest Service lost the public's high regard. It failed to embrace progressive environmental values, which broadened and deepened into a global phenomenon. The Forest Service became a target for the environmental community. The agency stood accused of selling out to commercial interests and squandering its legitimacy as a conservation leader— tragic, and generally true.

Famed ecologist Aldo Leopold worked for the Forest Service early in his illustrious career. Late in his life, Leopold wrote *A Sand County*

Almanac.[3] He humbly observed the woods and prairies of Wisconsin for a year, then drew on other important events in his life to create a larger understanding of the environment as a whole. He employed a keen eye and a pliable heart.

Leopold's stirring, sad account of his thoughtlessly shooting a wolf then watching a "fierce, green fire dying in her eyes" changed him forever. Leopold saw something new in those eyes—neither the wolf nor the mountain agreed with his simplistic notion that fewer wolves meant more deer, thus happier hunters. A new clarity illuminated the vital role of predators, and Leopold witnessed the pervasive damage from too-large deer herds.

Visiting the Colorado River Delta in Mexico, Leopold witnessed a wealth of fish and fowl and the pervasive presence of jaguar. He felt immersed in the Pleistocene. Leopold observed in the local activities of man an age-old cleavage: man as conqueror or man as biotic citizen. He said, "Man always kills the thing he loves, and so we the pioneers have killed our wilderness."

Leopold contended ultimately that we are part of our environment and must behave as biotic citizens. This, in turn, yields a land ethic that sees land not as mere soil to produce commercial crops, but as a biota of which we are an integral part, to be managed as a natural environment, not an artificial one. Ultimately, this is the challenge for the Forest Service and, more broadly, for all natural resource managers.

My story follows, describing how people, events, and places changed me from a naïve and optimistic young man to emerge from a crisis of conscience as a disappointed, more sober, yet hopeful architect of change. A friend once described the Forest Service I worked for as mostly uncritical lovers of the agency, yet surrounded by unloving critics. I hope to pursue the path of a critical lover, with deep affection for the Forest Service but convinced the agency must undertake profound change if it is to match its great calling.

Chapter 1: A North Woods Foundation

MAINE, 1968–1971

What's not to love about the de Havilland Beaver? The Beaver, a work-horse float plane common in lake country, turns almost any water body into a landing strip. I loved hearing but, even more, *feeling* the familiar rumble of the big nine-cylinder engine as we taxied to the end of a small no-name lake in Maine's North Woods.

Glen Sherman, the pilot, thrust the throttle lever full on, and we accelerated nicely toward liftoff speed. We'd done this numerous times, and it was always a thrill, but from my front passenger seat, the trees on the far shore were beginning to look, well...very tall, utterly insurmountable. Based on my limited experience, the odds that we would depart the lake in one piece were slipping to zero fast.

I really wanted to say something to Sherman, but I didn't think he would have heard a word. His eyes were locked on the trees, his jaw tight. There was nothing to be said.

Well past the point of a certain wreck, Sherman jerked the throttle back and we lurched to an abrupt, rocking stop just short of the shore-line. He looked me in the eye and said, quite calmly I thought, "One of you guys is gonna have to get out."

As Sherman taxied back to where we had started, Tom, my partner sitting in back, and I quietly weighed who would stay and who would fly. Was the margin for a crash only two hundred pounds? I didn't think it fair to ask Tom, a college student, to be the guinea pig. I, being the crew leader after all, stayed aboard. Tom got out.

His countenance composed and confident, Sherman gunned it one more time. Same acceleration, same looming trees, and a new lump in my throat. Then Sherman hit the flaps and we abruptly nosed up, up...and away. I simply said, "Nice job, Glen," as if he'd had it in the bag all along. A wink back.

Sherman ferried me over to a larger nearby lake where I waited, hopefully, for his and Tom's return. In a few minutes the silence was broken by the thrum of the Beaver. A routine landing, the pickup, and then we were off the water for the scenic flight back to the docks of Old Town.

That aborted takeoff is seared into my memory. When I see the once-proud Forest Service enmeshed in a largely self-made crisis, I think back and realize that Sherman had a choice: press on and try to lift the too-heavy plane off the lake—or stop and make a critical adjustment.

The Forest Service had blossomed after World War II, growing rapidly to provide timber resources essential to America's housing boom. The nation needed lumber and plywood, and the Forest Service met the challenge. The heavy logging produced a vast network of roads into wildlands, which had the added benefit of making these lands accessible for recreation enthusiasts. The Forest Service prided itself that TV's *Lassie* featured Cory Stuart as the iconic good ranger.

But by the 1990s, the Forest Service was in deep trouble. The northern spotted owl became the symbol of widespread gridlock, with loggers and tree-sitting environmentalists at each other's throats. Lawyers became the new gladiators. Forest Service morale hit rock bottom; public cynicism was high. An agency once lauded for its indomitable "can do" spirit became immobile and inert. The environment and the economy seemed two ends of a tightening vise squeezing the life out of the Forest Service. Paralyzed by litigation, the Forest Service dried up as a reliable provider of timber for a hungry industry. But neither could it gain the trust and respect of a cynical environmental community.

The Forest Service was no mere victim of circumstances. Its leaders had piloted the agency beyond the point where an adjustment could

have prevented a crash. Consider this: after decades of increasingly stri-
dent conflict about logging, more timber was harvested from national
forests in 1989 than ever before. And then in 1991, federal district
judge William Dwyer found that Forest Service officials had willfully
violated laws such as the Endangered Species Act in its pursuit of this
accomplishment.

The Forest Service embraces a great mission and proudly aspires to
public service. Utter darkness overwhelmed me when I read the federal
judge's words blaming people I knew for willfully violating environmental laws. How could the agency I loved have sunk to such depths?

That same year, Jack Ward Thomas, a wildlife biologist, became chief
of the Forest Service. He immediately issued a six-point directive that
included mandates to tell the truth and obey the law. What saddens
me is that the head of a once-trusted agency implicitly admitted that its
leaders lied and broke the law.

Like my pilot Sherman, Forest Service leaders must have surely
seen the approaching danger. Why did they not throttle back, stop, and
change their fateful course?

I will try to answer that question over the course of this book. Part
of the answer will be provided by relating and commenting on events
during my tenure as Siuslaw National Forest supervisor from 1992 to
1999 and deputy chief of the Forest Service from 1999 to 2002. In both
positions, I grappled with significant environmental controversies. I
strived to obey the law, and I told the truth. What a difference it made.

Today, the Siuslaw National Forest continues to be managed under
principles I established during my tenure. We throttled back, stopped,
and changed course, achieving a new balance that satisfies former environmental critics. Timber harvesting has returned following a nearly
total shutdown, yet with a difference. No more clear-cutting of old-
growth forests; instead, plantations of young trees are thinned to has-
ten the return of old-growth forests in the future. The Forest Service,
along with many partners, restores creeks and estuaries so that endangered salmon can begin the long climb back to their former abundance.
In addition, thousands of miles of unneeded logging roads have been
closed to improve wildlife habitat and water quality.

As deputy chief, I helped develop a policy to protect fifty-eight million acres of roadless areas in our national forests (2 percent of the land area in the United States) from further commercial logging and road building—one of the most important and most controversial actions in agency history. Roadless protection also succeeded as a popular public cause, yet, as I will detail, it fractured allegiances within the Forest Service.

I will share more about these and other experiences within the larger arc of Forest Service history, describing how the Forest Service and I both changed.

Roots

Almost everyone I've met in the Forest Service tells their career story with roots in the formative years of childhood events or in the influence of important people. I'm no different. For me, time spent in nature with my dad bent me toward a natural resource career. My father chose geology but was more fundamentally a devoted naturalist. His early career in petroleum took our family to Tyler, Texas, where I was born in 1945. When I was a year old, our family moved to Maracaibo, Venezuela, for three years, then to Dhahran, Saudi Arabia, for three more years, until Dad returned to his love of teaching and research at the University of Iowa, where he had earned his bachelor's, master's, and doctoral degrees during the Depression.

Dad studied nature with an inquisitive mind and a sharp eye, and he read voraciously. His experiences provided the best education, and he spent a great amount of time outdoors, constantly learning from the days of his youth until he died at age ninety-five. He strived to teach me what he knew about hunting, fishing, canoeing, birding, trees, geology, and many other things. Some of it stuck, and all of it affected my heart and my attitude toward our Earth.

He led the University of Iowa's summer geology field courses, spending six weeks in South Dakota's Black Hills followed by four more weeks in Wyoming's Bighorn Mountains. From about the age of eight, and over the course of several succeeding summers, I pretty much did whatever he did. I spent many hours atop an old ammo box in the front seat of his green Willys truck and probably more back in the pickup bed. Often, after completing the ten-week regimen, we'd take off on extended road trips all over the West as he consulted with graduate students on their fieldwork. I got my fill of trout fishing, big bucks and bull elk, soaring eagles, sage grouse for supper (I've had better meals), rattlesnakes, and—just once—observing a triple rainbow outside of Ranchester, Wyoming. I didn't know the word then, but heading back to Iowa for school at the end of summer was a downer.

The West infected my soul. It's hard for me to put in words how those summers with Dad shaped the trajectory of my life, but somehow I just knew I wanted to—had to—work outdoors and out West.

When I was fifteen, I was hired as a gofer and all-around handyman by the R Lazy S dude ranch in Moose, Wyoming. I was immature and unready and didn't fit in well with the college students. Lady Mac, the owner, fired me and sent me home, humiliated. But she told me she appreciated my love of nature and that I should follow my heart there. That helped soothe the sting.

As it happened, that same summer my older brother Dale was working in Quinault, Washington, on the Olympic National Forest. He and Larry Royer, a classmate at Grinnell College and a smokejumper, had dropped me off at the R Lazy S on their way West. Dale got the job at Quinault because Dad's cousin, Dick Swartzlender, worked on the Olympic. Dale brought back some great stories of working for the Forest Service. I revered the hobnailed boots he bought to get around in the steep, wet mountains of the Olympic Range. The boots seemed a talisman to me.

I'd always done well in mathematics, so when I entered college, I hatched a plan to major in civil engineering and get a job building roads and bridges with the Forest Service. This seemed simple enough. What I failed to consider was my dislike of engineering. Or was it my poor academic performance? Most likely these two factors were connected. I decided to take a more direct route and changed my major to forestry. My disposition and grades both improved.

My most potent memory of forestry at Iowa State came while strolling across the central quadrangle. George Thomson, a legendary professor nicknamed "Silvertip" because of his silver-gray hair and imperious, grizzly-like disposition, overtook me with his long strides. Matching my speed, he confronted me to inquire, "Do you want to be a forester or a swimmer?" I swam for the Cyclones as a determined but second-rate breaststroke specialist. We practiced from three to six every afternoon. This was a problem because Thomson taught my fall-trimester aerial photography class, which had Monday, Wednesday, and Friday labs scheduled from three to six. I fudged on both obligations, sometimes cutting class to swim, but more often missing swim practice. I'd pleaded academic hardship to Coach Jack McGuire and hoped I wouldn't get lost in his doghouse.

I had no such excuse for Thomson, but he awaited my reply and I told the truth—"I'm trying to do both." He said, "I'm not sure that will work for you," and then he picked up the pace to leave me alone with that thought.

I think I went to class more regularly after that encounter, but I also stayed with my swimming, and I'm glad I did. Our team was the champion of the Big 8 (as it was known then), edging Kansas. I took home some hardware for my fourth-place finish in the 200-yard breaststroke. Beyond that, I enjoyed the deeply satisfying experience of achieving my peak potential, being part of a winning team, and learning the benefits of hard work.

During college, I picked up a couple of great summer jobs with the Forest Service in Oregon. The first was at Tiller, on the Umpqua National Forest. John O. Wilson, an Iowa State graduate who knew my Dad, liked to hire Iowa boys. The next summer, Bud Sloan, another Iowa State graduate, hired me to do forest survey work on the Malheur National Forest in John Day. In that era, family and academic connections helped a great deal more than they do today.

Graduation day approached with job prospects still shaky. Ken Ware, one of my professors, mentioned that Joe Barnard, who worked in Forest Service research, would be on campus. I knew very little about the agency's research branch, the largest forest research organization in the world. Ware and Barnard shared work on forest statistical analysis, and Ware, knowing of my forest survey experience on the Malheur, encouraged Barnard to interview me for a couple of openings in Maine.

Barnard and I had a pleasant meeting. I still don't know exactly how it happened, but just before graduation, I received a job offer to join the Forest Survey group in Orono, Maine.

Maine Beckons

I came to work in Maine in May of 1968, having avoided Vietnam due to the birth of Julie, my first child.

Maine has lush forests, lakes to rival Minnesota's, and still-wild rivers like the Allagash and St. John. Forestry in Maine has a long history as a leading sector in the economy—from the earliest days, when loggers cut tall pines for spars in sailing fleets, to our modern era, with Maine's forests supplying paper pulp, fine veneers, and lumber. Almost all of the state is forest covered, and most of that forest is privately owned. When I arrived in 1968, five timber companies owned over 50 percent of the entire state!

I immersed myself in my new job, faced with the challenge of trying to make ends meet for a young family on $5,565 per year. I spent three years traveling the state engaged in forest inventory work, systematically sampling forest sites to provide an estimate of forest conditions and trends. At times the air seemed to consist of nothing but black flies and mosquitoes. No work boots I found could withstand the ever-present water. As time passed, I began to develop a feel for Maine, which struck me as static. The North Woods seemed deep, remote, and in my mind would eternally remain so. The people struck me as independent, conservative, and resistant to change. They liked Maine the way it was.

Yet Maine also offered the delights of fresh-boiled lobster, rocky surf, and flame-orange sugar maple foliage in October. The thick, quirky Maine accent, while hard to decipher, proved charming. Through it all, "Mainiacs" remained aloof and cool to newcomers like me, whom they described simply as "from away." Beyond Bangor and the stench of the big paper plant at Old Town, the pleasures of fresh air and nonstop woods greeted us on our trips to the North Country via Interstate 95, coursing all the way to Houlton on the border with New Brunswick, Canada.

Nothing prepared me for Maine's industrial-strength logging. Axmen had long ago stripped towering white pines out of the woods for spars on sailing ships. During the fur-trading era, hemlocks were felled merely for the tannin in their bark, used to cure the furs. Today, big sugar maples and yellow birch head to market as high-quality veneer logs. While lumber was the staple of the industry, niche factories cranked out toothpicks, matches, coat hangers, and wooden toy parts.

Logging operations combed through the woods, cutting the most valuable trees and leaving the rest—a practice called high grading. In many cases, clear-cut logging took it all, leaving the land to recover as best it could. With abundant rain and Maine's stony soils, new forests grew quickly. But the resulting mix of species could be radically different, and the new forest seldom resembled the old. These practices systematically converted older forests to young forests and also harmed other values like wildlife habitat and clean water.

On a site east of Moosehead Lake, not far from Kokadjo, I first saw a new forest harvester at work logging the flat terrain. A harvester is a one-man machine that navigates the woods on tracked wheels, like a tank, and then reaches out with a boom to grab a tree while a hefty saw severs it from the stump. The boom then picks up the tree like a big stick and sets it on a pile with others. The harvester can even strip off all the limbs and cut the tree to specific lengths. Several hundred acres of this site had been effortlessly mown down by this marvel of efficiency.

We walked about where the harvester had felled the spruce woods the previous year and found an enormous, unbroken landscape of raspberry shrubs. I honestly could not conceive how this site would ever return to forest by itself. I'm sure the timber company made good money on this logging operation, but none of it was invested to replant the site.

Little care was taken in constructing roads, and when the logging was done, most roads had simply been abandoned to bleed sediment into the nearest stream. These forestry practices didn't approach management as a means to invest in the land. This was all take.

I didn't have to sit on a board of directors or in a head forester's office to see an iron principle at work in the Maine woods—money counts.

I had little experience or perspective then. My forestry education had focused on the practice of forestry as "doing what's right." During my summer jobs working in Oregon on the Umpqua and Malheur National Forests, I mostly heard about the pursuit of "good forestry." Obviously, logging was also big business in Oregon, yet I didn't recall money being the policy driver. I heard discussions about doing logging, road building, and tree planting "the right way." The US Forest Service introduced me to taking pride in caring for the woods. Money was seldom discussed.

Emerging Concerns

To my young eyes and ears there existed a sharp difference between the private-land forestry I witnessed in Maine and the public-land forestry I'd seen in Oregon. In time, I came to discover two things: money does indeed govern much federal forest management policy, and private forestry devotes itself more to caring for the land than I once believed. But as a general axiom, money talks loudest on private forests, while the public demands and receives better attention to the environment on public lands.

Great Northern Paper Company, headquartered in Millinocket, was rumored to have logged only about one-third of the nearly five million acres it owned by 1970, with much of that ready to log again. I saw no reason to doubt it, as I had traveled over much of its land. When I left Maine, I took with me the impression of Great Northern's operation as a vast industry-owned commercial forest poised to be managed well into the future, as it had been in the past. Timber was king in Maine and would remain so.

Or so I thought. But events often mock assumptions. As it happened, before the twenty-first century arrived, industrial forest holdings across Maine would be bought and sold again and again. Their true value was no longer based on forestry and timber uses, but primarily as a mere capital asset: real estate. Yet, logging continues to be big business in Maine.

Though Maine seemed ever-wedded to industrial logging, it was far-sighted enough to create Baxter State Park before all the wilderness disappeared. Baxter stuck out on Maine maps like a green thumb, akin to New York City's Central Park. This park, named for former Governor

Percival Baxter, occupies over two hundred thousand acres (about 1 percent of the state) northwest of Millinocket in the heart of Maine's North Woods, and it exists, literally and figuratively, as an evocative counterpoint to private, industrial forestry. Dedicated primarily to recreation and preservation (a "Scientific Forest Management Area" portion is certified by the Forest Stewardship Council as a showplace for sound forestry), it represents pursuing forest values and objectives other than timber.

Maine folk are enormously proud of Baxter State Park, in large part because it retains the wildness and character forsaken in much of the rest of the state. The beaches and harbors of Acadia National Park and eating red, fresh-boiled lobster at a seaside "pound" have drawn people to the coast for decades. A "Maine vacation" is a cherished tradition for many Easterners. But a new spirit of adventure is afoot.

Whitewater boating did not exist when I worked in Maine in the late 1960s, with the exception of canoeing the Allagash and St. John. But now recreation booms on many of Maine's big rivers. Sailing and sea kayaking, as well as bicycling, enjoy increasing popularity. The ubiquitous cabin on a Maine lake makes a quiet but persistent claim for recreation as the beating heart of the Maine woods.

The notion of recreation usually implies simply having fun, but, taken to a deeper level, the meaning and practice of re-creation has a deeper spiritual connotation. Especially in natural environments, humans seek to renew and revitalize a connection with creation. We yearn for beauty, peace, and quiet. Our shared experiences remind us that we are but small travelers who live in a grand world that can function even without us.

How can I, or anyone, fully articulate the experience of sitting seaward on Mt. Desert Island's pink-gray granite headlands while an October full moon rises above the eastern horizon, the surf rhythmically surging against land? It would take a cold heart not to be moved and a fool to say anything at moments like that.

When most people picture themselves camping in a forest, they think of natural, unspoiled woods. I think about going back to a favored place, perhaps a childhood camping spot with fresh-caught brook trout sizzling in the skillet over a bed of coals. I'd be shocked to find it logged over. A common reaction would be disappointment and heartache at least, maybe outright anger.

The forest battle between commerce and protection is an old one. In the Maine I saw in 1968, with the exception of dedicated parks like Baxter, the scale tipped strongly, nearly exclusively, toward commerce on private timber lands and thus to heavy logging. Yet in the latter part of the twentieth century, when these timberlands had attained greater value as a real estate asset than as a wood factory, millions of acres were bought and sold repeatedly. Such boardroom decisions reflect a greater complexity and ambiguity about the values driving forest management—certainly more than mere timber.

Going West, At Last

We completed our inventory of Maine's forest resources after three years. I'd spent my time observing, learning, doing the best I could at my job. Although I developed some affection for Maine, I had no desire to stay for the next assignment in either Massachusetts or New Jersey. I yearned for the West, recalling my disappointment about my first job after college being out East. I should have considered myself fortunate to have been offered any job at all, since those who graduated after me struggled to find forestry jobs of any kind.

Seniority governed transfers of our Maine staff—the anointed one at the head of the queue would miraculously be placed in a job somewhere. I found myself edging toward the top of the list in 1971 when Jack Peters, my boss, told me the Black Hills National Forest in South Dakota hired two foresters. Paul Ruder, fresh out of Vietnam, went to Custer City. My heart and I headed for Deadwood.

How wonderful, I thought—blessed to be going to the same Black Hills where I had spent youthful summers tagging along with my dad as he conducted a geology field camp for the University of Iowa. Surely, this was the stuff of dreams.

Chapter 2: A Dream Realized

BLACK HILLS NATIONAL FOREST, 1971–1976

We made the long trip from Maine to the Black Hills in our VW station wagon and arrived in Deadwood, South Dakota, in July of 1971. My sharpest memory of Deadwood dated from early September 1957, en route to Iowa with my dad from our adventures in the West. The Willys pickup's radio had crackled with news of a big fire that nearly burned Deadwood down. The following day, as we passed through the eerie gloom of the blackened forest, I thought the highway itself smoked. As we arrived fourteen years later, the old burn was a sea of grass and dead snags and young planted pines.

We settled into an old rental house in the labyrinth of roads running uphill from the historic Franklin Hotel. I met my new boss, John Baswell, who explained that, as a trainee forester, I would need to learn about everything that makes a ranger district work. My first in a variety of jobs was on the timber-marking crew, led by Guy Virkula, a local Finn who grew up in nearby Lead, once home to the Homestake Gold Mine. Virkula, a forestry technician with a wealth of experience, showed me the ropes.

I spent almost all my time in the woods. As the months went by, Baswell continued to challenge me with new assignments and gave me encouragement along the way. "We'll make a forester of you yet!"

Water, Water, Everywhere

A group of Forest Service staff walked back to our vehicles after a timber sale review on Friday afternoon, June 9, 1972. We noticed a huge thunderhead that seemed to occupy the entire western horizon. I'd never seen anything like it. I felt the first rain on my shoulders as I walked home for dinner in Deadwood. I decided to unwind that night at the famous Old Style Saloon No. 10. I recall standing in the doorway of the Old Style about ten pm, beer in hand, saying to Harley, Deadwood's summertime community band director, "I've never seen such hard rain and so much lightning."

Saturday morning came. Indulging myself, I slept in until about 8:30. When I turned on the radio, I heard the announcer say, "Fifty-eight dead in Rapid City." Startled, my attention riveted on the radio for hour after hour, I despaired as the dreadful death toll continued to climb, eventually reaching 238, with many hundreds more injured.

At least fifteen inches of rain had fallen in less than six hours—recording rain gauges had overflowed and couldn't collect any more water. A swollen Rapid Creek swept through a densely populated flood plain from about ten to eleven pm, trapping many people asleep in basement bedrooms. Others fleeing in cars were swept to their deaths by the rampaging creek.

A fellow Forest Service employee, Luther Pullins, shared with me his tale of being trapped in his car along Boxelder Creek west of Nemo where he had to spend a cold, wet night in the woods. What a show he witnessed—incessant lightning and pummeling rain. Luther described the deafening noise of the raging creek and the sound of boulders rumbling—*ka-whunk, ka-whunk*—along the stream bottom. As flood waters rose, trees mobilized and migrated until thousands of trees, roots and all, washed out bridges and accumulated in huge debris islands, often several acres in expanse.

Boxelder Creek, normally ten feet wide and eighteen inches deep, pinches into a narrow canyon near Nemo. As I made my way down the highway to this spot a few days after the flood, I noticed a large pine tree perched on the banks of the creek. Its bark had been peeled off forty-three feet high by the flood. I tried to contemplate what this could possibly have been like at peak flood. Upstream at the Boxelder Job Corps Center, I saw a pickup truck with its tires perched on six-inch pedestals of gravel. The rest of the gravel had been swept away by floodwater sheeting across the parking lot.

As I drove from Rapid City to Sturgis a few months after the flood, I saw to the west a massive alluvial outwash from Little Elk Canyon, a narrow canyon where the road had been completely obliterated. A fresh array of one- to two-foot-thick boulders fanned out across the ground as much as a quarter-mile from the canyon's mouth. Then, to the east, I noticed even larger boulders from a long-ago flood that had come to rest on the other side of the freeway. I gained a new respect and appreciation for how periodic big floods do the work of carving canyons.

I thought about what I'd seen and what the flood revealed: Ease of construction leads to roads built in flood plains. Floods move massive amounts of water and debris, and the combination of the two magnifies a flood's destructive power—bridges and culverts pass water easily but not debris, leaving them highly vulnerable to destruction in floods. Human engineering tends to badly underestimate the size and frequency of natural phenomena—five-hundred-year floods seem to happen more often than predicted.

The result was massive losses to our road and bridge infrastructure. I'd noticed logging roads in nearly every nook and cranny of the Black Hills, which required a tremendous investment and high annual maintenance costs and placed a heavy environmental footprint on the landscape. While roads have asset value, the big flood exposed vulnerabilities leading to costly, yet predictable, infrastructure losses.

Ponderosa Country

George Armstrong Custer led an expedition to the Black Hills in 1874, about two years before he was killed at the Battle of Little Bighorn. William H. Illingworth, a photographer, was among the more than one thousand in the party. Illingworth laboriously chronicled the expedition, but the only remnant of his labors are seventy-nine glass-plate negatives that eventually found a home with the South Dakota State Historical Society. There they lay in the original wooden boxes until "discovered" nearly a century later. The Historical Society and Forest Service seized the opportunity to take new photographs at the same point as the originals. This effort resulted in a 1976 book, *Yellow Ore, Yellow Hair, Yellow Pine,*[4] which depicted ecological changes over the previous century.

The book proved to be a gold mine for naturalists. Comparison images one hundred years apart made it immediately apparent that the ponderosa pine forests of the Black Hills had changed dramatically. The 1874 forests reflected an active fire history with patchy forests having an open, park-like character, with trees both large and small strewn about the landscape—forested, yes, but not heavily. Thunderstorms pepper the Black Hills during the summer, capable of igniting numerous fires. In a natural regime with no means to extinguish fires except occasional rains, these fires created a mosaic of grassy meadows and forests with large old pines whose thick bark allowed survival during cool, running ground fires that consumed primarily grass. Such fires killed many, but not all, young pines. This represented classic ponderosa pine fire ecology. Research shows that most locales burned every five to seven years.

After publication of *Yellow Ore, Yellow Hair, Yellow Pine,* I was selected to make presentations of "The Black Hills—Then and Now" to schools, service clubs, and community groups. The interpretive slide show presented side-by-side images of the Black Hills in Custer's day and in 1974 to create a fascinating visual journey through familiar landscapes. Historical anecdotes of the expedition, such as excerpts from Custer's letters to his wife, made for a great program. Appearing in uniform, I enthusiastically sold the Forest Service via the scripted presentation—both the agency's interpretation of the ecological changes and its approach to managing the forest. We strived to tell the truth as we saw it. Yet, looking back, I now regard the truth as more complex and nuanced than I did then.

The young Forest Service aggressively suppressed fire to "save" the Black Hills forest created by Teddy Roosevelt, whose image is carved in the Mt. Rushmore granite. By excluding fire, the Forest Service set in motion a slow but inexorable transformation of the Black Hills, as could be seen clearly in the 1974 photographs. The forest we managed in the 1970s resulted from far fewer fires, and smaller trees survived in large numbers to become dense, young forests. Decades of logging had focused on removing the largest and most valuable trees. As a result, the Black Hills, when their dense forests were seen from a distance, became blacker than ever. All the while, the forest became less diverse.

In my slide presentation, I made much of how fire dominated the landscape. The forest was "saved" by thorough and systematic suppression of fires. Saved, perhaps, yet the hundred-year-old images also revealed how dramatically the Black Hills had changed due to persistent fire exclusion. Illingworth's images hinted that perhaps all was not right.

The "Big Blowup" of 1910 in Idaho, Montana, and Washington killed many people and leveled entire towns to ash while burning more than three million acres. The Forest Service responded with a ten-by-ten policy—all fires were to be held to less than ten acres, and they were to be extinguished by ten o'clock the morning following initial attack. In spite of the objective to keep fires small, many fires escaped to blow up to huge size. This remains true even today despite tremendous increases in firefighting technology and expense. We now know that, while infrequent, huge fires such as the Yellowstone National Park fires in 1988 are definitely normal. However, systematic application of the ten-by-ten policy throughout the twentieth century over most of the mountainous West prevented many large fires and unquestionably altered this larger landscape.

The Black Hills provide a small-scale example of a trend evident throughout the West. Wildfires now burn bigger and hotter. Ever larger numbers of people live in and near the woods, creating new threats to life and property. Costs of fire control have grown exponentially. By keeping fire out of the forest, we have unwittingly made forest fires much more risky, costly, and uncontrollable.

In those days, my views reflecting the Forest Service I worked for, I believed fire suppression was good, always and everywhere. Now I know it isn't that simple.

Also in 1976, shortly after new photos were being taken of the sites in Illingworth's images, I inspected a proposed timber harvest area very near the route used by the Custer party as they traveled through the Black Hills, just south of where Trout Haven stands today. I drove on an abandoned section of US Highway 385, the main north-south highway through the Black Hills, until the cracked and platy pavement yielded to quite a few young pines that grew up through the crumbling roadway, making further progress impossible.

On the road's shoulder I noticed a huge, beautifully weathered sign suspended from a hand-hewn log hanger. The sign had been crafted by the Civilian Conservation Corps in the 1930s. Its routed lettering read:

Timber Sale Area
Timber was harvested from this area
in 1905. A second crop will be
ready for harvesting in the near future.

I sensed history talking to me. I knew the sign referred to Case One, the first commercial sale of timber from national forest lands in the entire United States.

When I returned to the office that afternoon I fished the timber atlas out of the files. I located where I'd been on the map, and there it was—a timber sale area colored in delicate pastel, dated 1938; and underneath that, another color denoting Case One from 1905!

Shortly after I found the sign, the area was logged about 1976 for a third time, according to the plans I'd made. I visited the site in 1995, and saw fresh evidence of a fourth logging in only ninety years.

Based on the stumps I observed, the 1938 sale had removed most of the remaining big trees in the area. In 1976, we logged the last, scraggly big pines and then thinned the young trees. The abundance and size of the young trees made for good pulp and post material. The remaining well-spaced forest of young, healthy pines sat poised to make any forester's heart proud. Indeed, in 1995 I viewed a beautiful forest. Yet this forest bore almost no resemblance to what grew there naturally, except that the trees were the same species.

I'd seen a similar transformation throughout virtually all the Black Hills owing to many decades of logging. Big old trees have the highest value for lumber, and so these were systematically harvested. Remaining young pines were thinned to maximize future timber production. The remaining forests became nearly pure ponderosa pine but were stripped of virtually all their naturalness and diversity.

Most of the early logs departed the woods by rail. On maps, I could trace remnants of rail lines that snaked through valleys and draws throughout the Hills. As trucking became common, roads replaced rails to move logs to nearby sawmills. The rolling country of the Black Hills made for easy road construction, and a vast network of woods roads was created.

In one timber sale after another, I saw the same pattern in the Black Hills: rails replaced by primitive roads that were subsequently abandoned to build bigger and better roads necessary for bigger trucks. Of course, these new roads required gravel and gravel pits. Bigger and more powerful construction equipment also allowed more cut and fill to improve road alignment for faster travel. Not only had fire suppression and logging changed vegetation dramatically, but we'd created a landscape laced with both old and new roads.

Relatively low elevation, ample rain, deep soils, and reliable natural seeding for young trees make the Black Hills ideal for growing timber. A lively and competitive timber industry generated good revenue from timber sales. In short, forestry in the Black Hills made good sense, and the Forest Service delivered the goods. Fire-suppression efforts rarely let fires escape, timber sales provided predictable supplies to local mills, and the Forest Service road-building machinery performed with efficiency. Money from logging rolled in. A proud agency did good work.

But were there essential truths about the ecology of the Black Hills that deserved deeper respect? Although it didn't occur to me then, I might have wondered whether we could manage forests in ways that replicated ecosystems, sustained their function, and accommodated important phenomena like fire and flood.

Winds of Change

About 1975, Jim Hagemeier arrived as the staff officer in charge of the Black Hills timber program. A landscape architect, Hagemeier was among the first few nonforesters to crack the ranks of district ranger. Hagemeier had been handpicked by someone, but likely not the forest supervisor. Sending a landscape architect to the Black Hills to run the timber program was strongly symbolic—at best, provocative; at worst, simply dumb. I heard many foresters grumbling, jaws tightened with resentment, about how wrong it was to send us a "know-nothing."

From day one, Hagemeier let it be known that he pursued a new mission. He made it clear that he had his eye on a beautiful stretch of US Highway 385 just south of Pactola Reservoir that meanders through a gorgeous creek bottom with tall grass and pretty "yellow bellies"— big, mature pines whose bark had aged to a sublime yellow-orange that was beautiful from afar. Get real close to a big pine and you can smell the aroma—a spicy, vanilla-esque scent. It has thick, fire-resistant bark that easily flakes off in your hand, each piece unique as a snowflake. People like ponderosa pine forests—they are simply beautiful at an elemental level.

Until Hagemeier arrived, these big old Black Hills pines had a well-known destination: a nearby sawmill. Ponderosa pine lumber has strength, easy workability, rich grain, occasional knots to give it character, and, thus, very high value. There exists an implicit agreement between foresters and trees (even though trees have no say) —when a tree gets to a certain size, well... it's had a good long life, plenty of kids, only gonna be downhill from here, so "Adios!"

But Hagemeier insisted that these beauties stay put. He believed people drove this green tunnel for pleasure, and he knew they enjoyed these big, beautiful trees just the way they were. Here, along a major tourist highway, roadside scenery, not lumber, provided the highest value. He wanted these forests to retain their natural character, not become yet another forest landscape of cookie-cutter homogeneity.

For any forester schooled in agency dogma, this meant war. Bureaucratic war, anyway. Foresters worked at maximizing timber production, minimizing cost, designing the best logging practices, ensuring a fair price at sale, and overseeing logging operations toughly but fairly. You log it right, and people will like what they see—or at least you explain to them that they should like it even if they don't.

Scenery, driving for pleasure, and aesthetic values had no legitimacy to most foresters, compared to the value of forests as wood products. At stake was the right way, the tried-and-true traditional way, to practice forestry in the Hills.

Some would argue I'm exaggerating. Maybe, but only a little. I lived it and I could feel screws turning. I heard Hagemeier argue his cause passionately and persistently. I heard the reaction, both to his face and behind his back, and it reflected bitterness, resentment, and more than a little fear. I heard that "Hagemeier was screwing up a good thing," or words to that effect. "Hell, if it ain't broke, don't fix it."

Some heretics actually agreed with Hagemeier. Typically these were nonforesters such as hydrologists, wildlife biologists, soil scientists, and, of course, other landscape architects. Some foresters, like me, actually thought he made sense and that maybe his approach merited a try.

He and others like him were beginning to question the agency's fundamental values—whether the Forest Service was placing too great a priority on timber production to the exclusion or neglect of other important forest values. These internal voices were reflective of a growing storm outside the Forest Service.

By about 1970, the year of the first Earth Day, environmental activists in towns dotted throughout the Black Hills were becoming outspoken in their opposition to logging and roads. I recall Dave Miller in Deadwood, who argued that logging and roads were ruining white-tail deer habitat. Miller earned a reputation among our office staff as a loudmouth, a pest. Randy Fredriksen, an aide to Senator George McGovern (and who became a close friend of mine), related the "difficult questions" he was hearing from constituents about negative consequences of logging. I had no ready answers.

Implications

Near Roubaix Lake about 1972, I tracked through fresh snow on a mild February day under a bright blue cloudless sky. Amid a fresh logging operation, I heard the agitated cry of a hawk. A goshawk? I could see it now; yes, surely. I'd had a couple encounters with goshawks in Maine. Goshawks become incredibly aggressive when nesting, as this one was, and were known to swoop down and strike people. Homing in on the cries, I detected a nest about 40 feet up in a tree. The hawk boiled out of the nest and immediately dove at me. Ducking behind a tree, I heard the "whoosh" as it swept past, and I felt a whack as the wingtip struck my shoulder. The goshawk wheeled and gathered itself for another pass as I quickly left the nest site, and it continued to scream until I was gone.

What to do? Even though we had no goshawk rules per se, the answer seemed clear—suspend logging in the vicinity of the tree until the young had flown the nest. I spoke to the loggers and said I'd flag the nest tree later to make sure it remained standing for possible use in the future. A call to Fred Wild, the forest wildlife biologist, confirmed the plan, and I penned a quick note to the John Hallinan, the logging representative from Whitewood Post and Pole. Wild thanked me for making room for wildlife. He had the forest supervisor send me a letter (no doubt Wild wrote it himself) applauding my concern. That felt good—do the right thing, and sometimes people notice. Little did I know protecting goshawks would later become a critical legal issue that constrained logging all over the Black Hills.

My next assignment took me to Hot Sulphur Springs, Colorado, in the Middle Park area west of Rocky Mountain National Park. I worked for Grand County's Department of Planning, helping to develop a land management plan for the Grand Lake area. I took the opportunity because Sid Hanks, then deputy regional forester, dangled the prospect of a district ranger's job in front of me if I did well. Lo and behold, a year and a half later, I landed a job with the Bighorn National Forest in Wyoming as the Tensleep district ranger. It seemed too good to be true. I had become a ranger on the same national forest that had earlier in life won my heart to the West.

As I moved on to the next phase of my career, I took with me a realization that two major drivers dominated Forest Service activity— suppress fires and harvest timber. The agency budget and organization structure consistently reflected this reality. This worked itself out on the land through logging, roads, and keeping fire at bay. That is what the Forest Service did, and after several decades, our work dramatically changed the forest from what it had once been.

Yet the Black Hills, being South Dakota's only national forest, holds a cherished place in the state's reputation for recreation, tourism, hunting, fishing, and beauty. After my five years here, I sensed these to be the true values of the Black Hills. The iconic Mt. Rushmore, Wind Cave National Park, Custer State Park, the emerging Crazy Horse sculpture by Korczak Ziolkowski, and even Custer City's kitschy Flintstones Village and other tourist attractions draw millions of people to the Black Hills. But the forest itself, standing in stark contrast to hundreds of miles of surrounding plains, is the underlying fabric that holds things together and makes the Black Hills special. I thought Jim Hagemeier, the landscape architect, was right. Forest management must be designed with certain wisdom to maintain and foster nature's primary, basic values.

Chapter 3: A Rookie Ranger

BIGHORN NATIONAL FOREST, 1977–1984

"Adventures in Moving." The enormity of this lie on the back of the U-Haul truck reached a criminal level at the moment my wife, Judy, made an abrupt U-turn in her little Toyota Tercel. The disappearing taillights seemed to me the last straw on a hectic, tiring day leaving Colorado for Wyoming. I searched for a place to turn the big truck around, hampered by our second car in tow off the back. I journeyed all the way around the block since backing up was out of the question, and then...where to look for her?

Besides, I was still stewing over her "No deal" on the first two motels we looked at. Why did we have to bargain-shop for the best deal in Rawlins, Wyoming, if Uncle Sam was picking up the tab?

I finally saw her car in a motel parking lot after a few minutes of hunting around town. I located her in the lobby, receiving a key from the desk attendant. I found a place to park the rig and made my way to our room. A stony silence engulfed the small room. Let's just say it helped that we'd only been married five months and were still all aglow with love. I don't think she noticed I wasn't speaking to her until the next morning. I never did learn to balance the excitement of a new assignment with the stress of a move; it seemed to bring out the beast in me.

My first wife and I divorced in 1973 while we were living in Rapid City, South Dakota. Three beautiful young, innocent girls were the best part of our history. I took a position in Custer City at the forest headquarters for a couple years doing forest planning, then accepted a temporary assignment with Grand County at the county seat in Hot Sulphur Springs, Colorado. In October 1976, in Granby, Colorado, I met Judy, an Elementary school teacher, on a blind date—dinner at Bear's Lair and then a movie, *One Flew Over the Cuckoo's Nest*—and we married in Grand Lake a few months later.

We left Granby that July morning in 1977 after buttoning up the moving van and made it as far as Rawlins. It seemed like a good city for the State Penitentiary. Moving to Worland, Wyoming, had been such a happy thought just a day ago. I tried to calm my nerves while I mulled over the stress of moving. My "marriage" to the Forest Service required occasional transfers; each initially difficult, with many adjustments until normalcy and better times emerged.

Almost Heaven

Here, however, my good fortune seemed to have no limit. Selected as Tensleep district ranger on the Bighorn National Forest, I had an opportunity to serve on yet a second national forest, the very one where I had spent joyous summers as a youth following my father around on his University of Iowa geology field courses. Technically speaking, I lived in Iowa City, but my nine-year-old heart had taken flight to Wyoming by the time that first Bighorn summer ended. Now an adult, I was to discover trouble in paradise.

The first hint of trouble involved buying our house. John Puett, our realtor, helped us pick our dream house from among the very few homes for sale in town. The turquoise cinder-block ranch house had been vacant for over two years but had just been listed for sale. The owner had difficulty letting go of her childhood home after the death of her mother. Two-year-old food remained in the refrigerator. Personal effects hung in closets and lay in dresser drawers exactly as they had when the mother had lived there. We liked the house; it had charm. Hasty negotiations ensued with the husband of the still-grieving daughter while she was out of town. We closed the deal swiftly, though it almost came apart when she faced the need to sign away the house forever. We settled in to our new life.

Puett seemed quite knowledgeable about Forest Service affairs as we got acquainted, and he speculated that my administrative assistant at the office "would be gone within six months" because of a rumored romance with the previous ranger. In small Forest Service offices, the administrative assistant had a well-deserved reputation for being the ranger's right arm, the glue that kept the office running smoothly. She left in about five months, much to my chagrin, succeeded by six others in the span of my seven-year tenure at Tensleep. One lasted less than a day. I endured incessant jokes about my skills in finding and keeping good help. Puett's intuition had proved correct. My intuition needed work. My lack of experience created some obstacles down the road.

John Puett's knowledge of the Bighorn National Forest owed to frequent flights there in his tiny Citabria two-seater. He invited me out for a dawn show-me trip up on the mountain. Puett and I took several such trips that summer before he perished in a fiery crash while scouting elk herds in October. The news of Puett's death stabbed my heart and shocked the people of Worland, where he was well-known and highly regarded. I couldn't believe he was gone. Puett was a good man, a good friend, and he lovingly shared his intimate knowledge of the Bighorns. I imagined it could easily have been me on that fateful plane trip. I deeply admired Puett's passion for The Mountain, as locals referred to the beautiful Bighorns.

Looks can be deceiving. Truly, they can. The Bighorns stunned me as we flew low and slow in the clear air and early morning light. Cloud Peak, at 13,167 feet, sat high on the east horizon, herds of elk dotted mountain meadows, alpine lakes shimmered crystal blue (Mistymoon Lake won my personal best-name category), and Tensleep Canyon's goldish dolomite rimrock took me back to days with Dad naming rock formations.

At one point, Puett remarked on some cattle below not being in the right pasture. I found it a bit odd that he would know and care about such things. I was to discover that he, like many locals, truly believed that The Mountain belonged to him. As public land, it did—literally. I came to learn that there are John Puetts in communities all over America. I wondered what Teddy Roosevelt and Gifford Pinchot would make of the way people connected with public landscapes at such a visceral level. They might not be surprised. This offered the best evidence of their wisdom and vision when they created the first national forests.

As pretty as the Bighorn National Forest was, I erred in thinking this translated to its being healthy as well.

Ron Higgs, the district range conservationist, managed the grazing program. Higgs took me out for my first on-the-ground look around. We peeled off the graveled Battle Park Road, heading north on a rough dirt two-track from Brokenback Mountain (no doubt the inspiration for E. Annie Proulx's short story "Brokeback Mountain"[5]) toward Bald Ridge. When we crested a hill, Cloud Peak and the granitic crest of the Bighorns loomed in the distance, and a sea of emerald grass stretched in front of us. Not another soul in sight, and I felt like King of the World.

We were in a sea of grass—but there's grass and then there's healthy range, and they are two different things entirely. I had no clue what I saw before me. I was looking but not seeing. This mountain range was indeed pretty, but not in particularly good shape. Important characteristics like plant composition, density, and productivity were completely unknown to me. Regrettably, I didn't know what I didn't know.

I replayed what I did know from former lessons in Dad's geology field classes. The Bighorn Mountains display the character of a classic domal uplift. These are mountain massifs rising to over thirteen thousand feet in elevation from the relatively flat surrounding terrain, which sits at about four thousand to five thousand feet. Picture an egg, sunny-side up. The elevated core of the range displays heavily glaciated, craggy granite with hundreds of cobalt-blue alpine lakes. A mantle of surprisingly gentle terrain with a nice blend of pine and spruce-fir forests and open, grassy parks surrounds this granite core with abundant gray-green sagebrush and, in early summer, stunning wildflower displays. Long, heavy winters lead to a rush of spring snowmelt coursing through several spectacular canyons—Tensleep, Paintrock, Medicine

Lodge, Shell, Powder River, Tongue, Crazy Woman—that cut through the thick sedimentary shoulder of the mountain before emptying into the basin country.

As with almost all national forests, Native Americans once called the Bighorns home. Teepee rings alongside creeks in mountain meadows and lithic scatter from the creation of tools and arrow points still give poignant testimony to another time. Early homesteaders quickly filled the void created by the eviction of Indians and used the high country for summer feed for great numbers of sheep and cattle.

Iconic ranches, many originating in the late 1800s, snuggle up to the fringe of the Bighorns, with their cattle and sheep still driven to the high country for fattening up, as always. With livestock away from the home ranch during summer, irrigated pastures produce hay and alfalfa for harvest and storage to provision herds through the long winters. Thus, grazing on The Mountain has long been considered essential to most ranches in the area. With the creation of the Bighorn National Forest in 1897, the government handed out grazing privileges to ranchers in the form of permits to honor historic and traditional use. This custom applied throughout the West when forest rangers appeared and began to establish a new order.

I had the good fortune to work with men whose ranch culture formed the backbone of communities in the Bighorn Basin—Smilo Davis, Roy Frisbee, Clare Lyman, Gary and Ray Rice, Eldon and Jed Leithead, Bud Leavitt, Fred Feraud. Many of these Bighorn ranchers traced their grazing permit back several generations and saw their grazing of national forest range as a right, not a privilege. And there was a way things were done around here that I, as the new ranger, needed to understand. I quietly bristled at this attitude. These were public lands, not the ranchers', and the Forest Service's job was to ensure that grazing didn't harm the land.

Higgs and I were touring the Battle Park Allotment that July day. Several ranchers pooled their cattle—about fifteen hundred cows in all, along with a like number of calves—to graze this allotment in the northwest corner of Tensleep District en masse from late June to mid-October in a typical year. They pushed the big herd up and onto the forest through South Paintrock Creek, then grazed the herd in a leisurely counterclockwise pattern through Soldier Creek, Bellyache Flats, and finally the Buck Creek Vees, using up the grass as they went along, assisted by pasture fencing to hold livestock before moving on. They'd done this for many decades. But continued repetition of this pattern of grazing left a heavy footprint.

Grass enjoys a pretty simple lifestyle. When snow melts and days lengthen, it's time to grow. Fresh blades burst forth to capture the sun's energy through the magic of photosynthesis. Grass miraculously converts sunlight into food energy that trundles down into the roots for storage over winter until drawn on the following spring to grow a new set of leaves. Then the process begins anew.

This process takes time. The staple grass, Idaho fescue, so abundant in the Bighorns, will complete the job of storing root reserves by about August 20, give or take a few days depending on elevation and rainfall. But if you disrupt the plant's production capacity by eating the grass in July, for example, bad things happen. Rangeland research describes how excessive early grazing dramatically reduces the potential energy stored in roots. I arrived on the Bighorn quite ignorant of these basics and was fortunate to get some great schooling from Don Nelson, our range staff officer in Sheridan, Wyoming. Nelson helped open my eyes to the subtleties of what "good range" looked like. I needed to learn to see beyond simple instinctive reactions.

Idaho fescue should be common and robust for healthy range at this elevation. But in many places, abundant, mature sagebrush hogged soil moisture and severely diminished grass production. Thorough analysis of range condition required that we laboriously clip and weigh grass to gauge production and utilization in pounds per acre. We needed to methodically identify the exact species on numerous hundred-yard-long transects sampled every three feet so we could detect trends in range condition—was it improving or deteriorating?

The bottom line for plants is that grazing them early every year severely weakens their capacity to store food and dramatically reduces pounds per acre of grass produced. Plants grazed late every year, after having stored energy in their roots, remain healthy and productive, with high plant density and tall grass. As expected, portions of Battle Park allotment grazed early every year showed poor range conditions, while those portions grazed late were in good shape.

I came to realize that the Battle Park area required changes in the grazing pattern if we expected any improvement. This came to me slowly, but convincingly, based on irrefutable evidence—the grass told us what was going on. The tricky part for Higgs and me came in conveying this information to the ranchers, who viewed the rangers who came and went every few years with suspicion. Higgs and I had to help the ranchers over the same hump I'd faced in understanding why poor grazing patterns adversely affected grass and how good grass could add pounds to the calves trucked to market in the fall, resulting in higher profits. Ultimately, their willingness to trust me really mattered. Besides that, everybody needed patience to wait a few years for results.

About February 1979, Ron Higgs and I prepared for the annual spring meeting with the Battle Park grazing permittees. Higgs had done all the analytical work digesting our field data on grass production and range condition. This led to our carefully crafted proposal to institute a new grazing system that should improve range conditions (our responsibility) and also improve their bottom line of sending healthier, heavier calves to market (their goal).

Higgs and I made a great presentation in an academic sense—rational, well-argued, and practical. I thought anybody with a lick of sense would "get it" and enthusiastically rush up to sign on.

We didn't expect applause, but after we finished, an awkward silence hung in the room as we awaited a reaction. Most likely, a peer leader would speak first, but who might that be? Bud Leavitt seemed a likely prospect, or perhaps Ray Rice. We received a kind of group shrug, almost as if orchestrated in advance, followed by sentiments to the effect that, while this was all very interesting and we'd sure done our homework, they just couldn't quite see how all this would work out and they pretty much wanted to stick with the usual.

The drive back to Worland felt way too long as Higgs and I wondered out loud what it would take to get these guys to commit to making some changes. We were convinced we occupied the high ground while the ranchers stuck to their old ways, but we hadn't succeeded in breaking the mental inertia that necessarily preceded change. Should we consider something more potent than buttering them up with an optimistic future? Maybe we should cut their permit, reduce the number of cattle they could graze. That would get their attention! After the meeting, I gnashed my teeth and pondered various what-ifs. I felt as if a gray, cold fog had rolled in.

A couple of weeks later, one of the Battle Park ranchers, Ray Rice, stopped by the office. Rice said he'd come to town on a couple of errands and wondered if I had a few minutes to talk.

I didn't know Rice well, so this was a chance to get acquainted. I did know that he and his brother, Gary, had each inherited half their father Vernon's cattle ranch. Each half still amounted to a big cattle operation. The brothers had both excelled at football in Tensleep and exuded toughness. They used to host some of the Pittsburgh Steelers Super Bowl champions for elk hunting in the fall. Rumor had it that guys like Rocky Bleier and Franco Harris loved the Rice boys.

I already had a deep respect for Ray Rice. The previous summer I had accompanied the ranchers on a Battle Park tour with several other forest staff, talking cows and grass. When the young daughter of a rancher casually dropped her reins, her horse instantly bolted and took off—a runaway! But this was no movie. On a dead run, her horse headed for a barbed-wire fence about a quarter-mile away.

Rice sized up the situation immediately and, picturing the ugly scene when that horse hit the fence, spurred his sorrel hard. The girl did her best to stay aboard. Rice pounded leather to overtake her.

He slowly drew even, then as he eased in front, reached out with his left hand and grabbed the flying reins near the bit and abruptly "Whoa'd," bringing the runaway horse to an abrupt stop. We went from sleepy horse ride to impending disaster to event over in the space of about thirty seconds! As Rice brought the girl and her horse back to her relieved and grateful father, I remember a rancher, Roy Frisbee, saying to me, "Ray's the only horseman here good enough to pull that off."

I noticed on Rice's horse a neatly sliced four-inch pink gash behind each shoulder where his spurs had said, "NOW!" I silently thought about cowboys and some of the jokes I'd heard and told. Rice, right then, had stopped all that for me for good. The respect he earned in that moment was real and deep. I understood he had skills I never could appreciate fully.

So, when Rice asked if I had a few minutes to spare, I wanted to hear what he had to say.

He said he'd been thinking a lot about the meeting we'd had up at Hyattville and allowed as how he thought our proposal actually made some sense. He'd spoken with some of the others, and he thought they'd be willing to give it a try.

I was being schooled here. I didn't object. I actually enjoyed the sweet feeling that hopeful changes lay in store for Battle Park. Rice's methods differed from mine, but he'd delivered what had eluded me. Maybe they don't teach this stuff in school because it's too nuanced and particularly cultural. But to this day, I believe that Rice could not agree with me publicly in front of his people. He grew up with these men, played football and did rodeos together. His kids grew up with their kids. A ranger like me would be long gone in a few years. These ranchers were his partners, many for life. Yet I think Rice understood that the grass was our mutual asset. In a courageous and wise way, he said, "I'm in." Then he brought a critical mass of others with him, in his time, in his way.

This ranching culture runs thick and deep across the West. The relationship between ranchers and the Forest Service has evolved over a century. The first forest rangers wore guns. Some died from an unseen assassin's bullet.

Tension surrounds the issue of range ownership. Ranchers and their heirs brought cows and sheep to the same piece of high country, often for many decades. Although the Forest Service has authority to manage the land and carefully refers to ranchers as "permittees," these permittees know the land intimately. And so do their livestock!

Rangers come and go like flotsam on the beach. Come back in the morning, there will be new junk at the high-tide line. Even though these ranchers saw me as a visitor, Tensleep sure seemed like home to me. I fell in love with this country, hard. I wanted to believe that just being "the ranger" established my authority. But I had to earn respect and trust in ways that had nothing to do with my uniform. Being invited to a hearty breakfast at South Paintrock Cow Camp during the spring cattle drive was an honor, and attendance was obligatory. Besides, the food was damn good! Yet a fundamental divide remained—when the ranger decides it's time to make a change in grazing practices, ranchers weigh the suggestion from a perspective of what's good for the ranch, not necessarily what's good for the grass.

Two perspectives converge when what's seen as good for the grass is also seen as being good for the ranch. Grazing abuse of public lands is regrettably all too common, resulting in calls from some quarters that all grazing cease. But grasses evolved with grazing. Properly managed, domestic livestock grazing is a legitimate activity.

I've returned to Battle Park often. The ranchers? Some sold out, some have died, and some remain. We fired no silver bullet there. Our plan, though a significant improvement in my mind, fell short of perfection and had to be adjusted and tinkered with. Anger and feelings of betrayal followed when the Forest Service eventually reduced cattle numbers. I've come to believe that the only way to adequately protect most streams is to fence livestock out. I didn't require that when I had the reins as Tensleep ranger, and it still hasn't happened. Range conditions have improved substantially since 1980, and this trend needs to continue.

One person—Eldon Leithead—inherits much of the credit for the good things that happened. Leithead held a Battle Park grazing permit when I arrived in 1977, but soon thereafter he sold his ranch. However, the Battle Park ranchers quickly hired him back as their range rider, managing day-to-day affairs on the range—moving livestock onto fresh grass, putting out salt, fixing fence, taking care of business. This he has done faithfully for over thirty years. I saw him recently mounted on an ATV rather than a horse. Even cowboys can change!

I bump into Leithead repeatedly on return trips to Worland—almost like a touchstone—at church, the county fair, on the mountain. Our relationship remains respectful and warm. I feel a shared affection for a special piece of land that runs deep in his blood, much deeper than in

mine. Leithead quietly but assuredly did all he could to make our grazing plan work. Many range riders, hired by the ranchers, see their job
as just taking care of the cows. Leithead distinguished himself by taking
care of the land in his way, striving to meet both the ranchers' and the
Forest Service's goals.

Leithead chose to believe in a better way forward and committed
himself to help make it happen. He could have easily torpedoed everything if he hadn't diligently tried his very best to see that the allotment plan succeeded. I pinch myself for the good fortune that Leithead
shared our vision. I love that man. I love his faith.

Timber Country. Really?

When I travel through the Bighorns, the forests do not impress me
in a way that great forests of the Pacific Northwest do, with the sheer
gargantuan size of their trees, nor do they dominate the landscape in
the same way. Timberline—the elevation above which forests no longer grow—shows at about ninety-five hundred feet in the Bighorns.
Forests here also have a floor and grow no lower than about five thousand feet, limited by precipitation. Thus, they occupy a fairly narrow
elevation spectrum.

I say this to make the point that the Bighorns do not strike a person
with common sense as a place for timber production. However, when
the Forest Service committed itself to aggressive timber production after
the Second World War, evidence strongly suggests that virtually every
national forest, including the Bighorn, was expected to get with the
program and harvest timber at levels that aspired to maximum potential. How else could I explain seeing, over and over again, the pattern of
aggressive logging and road building on national forests aimed at taming these wild lands? Such strong evidence seemed the best indication
that the Forest Service created a mission and marshaled the agency's
resources to "get the cut out."

As a young boy in the mid-1950s, I don't recall seeing much logging in the Bighorns. Arriving as a young man and new district ranger
in 1977, I saw quite a lot. One of the first places I visited with district

forester Steve Uhles was a clear-cut area in heavy lodgepole pine timber just south of Powder River Pass, where it was proving enormously difficult to get planted seedlings to survive.

Ideally, the Forest Service shouldn't have to plant trees at all. Lodgepole pine reproduces extremely well, especially after fires. This fire-dependent species relies on cones that clutch the seeds until a fire's heat warms the resins, after which the cones expand and open. Later, after the fire has passed, seeds flutter to now-cool ground and—voilà! —germinate to establish a new forest. Fire kills the old forest but also creates a new one.

But here, standing near timberline at an elevation of about nine thousand feet in a clear-cut (not a fire), no trees stood ready to sprinkle down seed. Old lodgepole pine trees occupied this site about ten years earlier, but they became two-by-fours. The Forest Service had repeatedly planted now-dead pine seedlings. In doing so, the agency spent much more money than they ever received from Wyoming Sawmills in Sheridan, the company that logged the trees. How did we get here?

As Uhles and I stood in the barren old clear-cut, my mind wandered back in time. I've seen pictures of the "cut and run" logging devastation Teddy Roosevelt witnessed, which supplied much of the motive for the foundation of the conservation movement. When the US timber industry bumped into the Pacific Ocean, it had to become wiser and more mature. Far-sighted industrialists like Weyerhaeuser and Simpson elected to buy land in the Pacific Northwest and committed to logging and milling wood from their own lands, hopefully in perpetuity. These timber barons amassed great fortunes. Enormous profits could be realized so long as abundant, cheap supplies of federal timber from national forests did not flood the market, so the timber industry waged political battles to oppose logging of national forests. Thus did federal timber harvest remain relatively meager until the post–WWII housing boom.

When demand for wood products began to exceed private industry's supply capacity, strategists began to ogle the vast supplies on national forests. But few sawmills existed to handle all the federal wood. A critical need existed for greater industry milling capacity.

This issue contained an important subtlety. An expanding timber industry need not invest in land, only in manufacturing capacity. In fact, laws prohibit the Forest Service from selling land. Thus, a new segment of the timber industry began to expand rapidly. This industry niche sector owned no land, only mills, and necessarily relied on a supply of wood from national forests. As this industry sector blossomed, management of public lands responded rapidly with new roads and timber harvests aggressively pushing into previously untouched wild lands. Not incidentally, the Forest Service preferred clear-cutting high-volume, high-value old-growth forests to supply the industry's appetite.

The emergence of a logging industry in the 1950s and 1960s throughout much of the Rocky Mountain West resulted from a well-orchestrated collaboration between timber companies and the Forest Service. National forests had an abundant and cheap supply of trees to meet the demands for lumber and plywood from the supercharged housing boom. The Forest Service enticed the timber industry to build mills with de facto guarantees of a long-term timber supply from public lands. Industry then created the manufacturing base.

In *This House of Sky*[6], Ivan Doig describes how, in his youth, White Sulphur Springs, Montana, went from being a cowtown to a logging town. Log trucks increasingly frequented Main Street, and a constant smell of wood smoke hung in the air from the teepee burner that consumed the sawdust and trimmings. Hard hats outnumbered cowboy hats.

So Steve Uhles and I stood looking at the bare clear-cut in front of us, pondering this case in point and our particular predicament. The timber company Wyoming Sawmills built its mill in Sheridan, Wyoming, and for decades relied on logging Bighorn timber supplied by the Forest Service. This particular situation of reforestation-gone-wrong became common in the Bighorns after logging escalated rapidly in the 1960s, as it had on many national forests. Inevitably, many folks became concerned about the growing ugliness of leveled forests with alarmingly few of the promised new seedlings, and they galvanized for political action, as citizens often do when they get fed up.

In Wyoming, Senator Gale McGee (a Democrat; seemingly unusual for what's since become a Republican stronghold), intervened in the early 1970s and appointed a blue-ribbon commission to determine if national forest timberland was being mismanaged. Among specific concerns in Wyoming: why was the Powder River high country being clear-cut, and why were no new seedlings growing there?

The ensuing report from the commission offered a scathing indictment of Forest Service timber harvesting practices on the Bighorn and other Wyoming national forests. More than one head rolled in the aftermath. Cumulative political damage mounted from other similar crises in other parts of the country.

Two in particular proved to be pivotal in creating enough leverage to move political machinery. The Bolle Report, issued in 1970 by a group of academics headed by Arnold Bolle, dean of forestry at the University of Montana, harshly criticized logging practices on Montana's Bitterroot National Forest. An even greater controversy stemmed from an Izaak Walton League lawsuit, resulting in a 1973 judgment against the Forest Service practice of clear-cutting on West Virginia's Monongahela National Forest. The Bolle Report cast serious doubt on the professional competency of the Forest Service. The Monongahela judgment found that clear-cutting violated the requirement of the Organic Act of 1897[7] that trees be individually marked for selective logging.

When I arrived in Wyoming in 1977, federal forest management was in the midst of a seismic policy shift. The federal magistrate in the Monongahela lawsuit had struck down timber harvest practices that the Forest Service was doing virtually everywhere, threatening a nationwide shutdown of logging. Environmental groups, critical of the Forest Service but seemingly toothless, could taste blood. The logging industry, shaken and alarmed, called for swift, decisive political action to address the crisis. Forest Service leaders knew intervention was necessary to keep the timber machine from grinding to a halt.

Several congressmen, primarily senators from western states, had long supported the Forest Service's timber program—it sustained big business and the tens of thousands of jobs that depended on logging public land. In a bipartisan move viewed as nearly impossible today, the House and Senate swiftly swung into action in 1976 to enact the sweeping National Forest Management Act (NFMA)[8]. Hubert Humphrey, Minnesota's powerful senator, ramrodded the effort.

The NFMA sought to place timber harvesting in a broader context of forest management. It included numerous precautionary limitations, such as restricting clear-cuts to less than 40 acres, prohibiting timber harvest if reforestation could not be assured within five years, and requiring careful analysis to ensure that timber harvest did not exceed sustainable levels in the long run. The NFMA required a comprehensive management plan for every national forest.

Little of this history played in my mind as I stood there in the middle of the problem. I just focused on how to get young trees to grow—which, I might add, the Forest Service must do, as required by federal law. We tried. God knows we tried. Nearly every spring we planted as many trees as we could, depending on the availability of seedlings and available funds for contract tree planters. Other district rangers on the Bighorn often shared with me their lament over similar difficulties.

We ultimately achieved "adequate stocking" and declared victory, at great cost. How ironic is it that in many places lodgepole pine seedlings revegetate a site naturally and so thickly that you can barely walk through the woods? Today, some forty years after harvest, these scraggly Powder River saplings weakly resemble a new forest rising, if you use a bit of imagination.

It remained a painful lesson. It was possible to cut the stand of high-elevation lodgepole pine: laying out several clear-cuts, getting the roads built, and cutting the timber proved relatively easy. I wondered: should we have done so with full consideration of the possible consequences and costs? Clear-cuts at this high elevation, so close to timberline, evoked a quiet wrath from too many complicating factors—a very short growing season, thin soil, intense sun that scalded the seedlings, and a lack of tender loving care for seedlings at every step from nursery bed to the high mountains. Did we really need the two-by-fours that badly?

Situations just such as this stoked the fires of controversy leading to NFMA's requirement that the Forest Service develop a comprehensive land management plan for every national forest, ostensibly to foster sound forest management while preventing ill-advised actions. Thus began the era of forest planning.

Little "p" Politics

About 1980, the Bighorn National Forest undertook the challenge to craft a new plan as required by the NFMA. Welcome to the tempest.

Each forest plan provided a comprehensive, long-term evaluation and strategy for all resources. It quickly became clear that the crux of each plan was the timber component.

The Forest Service had developed quite good map-based, species-specific data showing acres, volumes, and age classes for the forest resource. K. Norman Johnson, then a forestry graduate student at Oregon State University in Corvallis, developed a powerful, complex computer modeling program called FORPLAN that enabled a planner to evaluate this kind of timber data to forecast the amount of timber that could be logged sustainably, in perpetuity.

FORPLAN analyzed data to derive estimated outcomes, necessarily relying on numerous assumptions. For example, if you clear-cut an area and replanted tree seedlings, how fast would they grow? Assuming growth rate x, FORPLAN estimated timber harvest yield at age sixty or one hundred. If one included monetary values, FORPLAN also estimated financial benefits and costs.

The Forest Service required that all national forests use FORPLAN in crafting their forest plans. However, it was complex, and few people understood it well. It quickly became the proverbial "black box." It seemed that the five members of our planning team were the only ones who could savvy what went on inside the computer. As a ranger, I understood the theory of FORPLAN but could not grasp how the model actually worked. All other staff were beyond their depth. We'd joined the Forest Service to work in the woods.

The Bighorn leadership group consisted of the forest supervisor, Jack Booth, his five staff officers, and five district rangers, including me. Booth had somehow pulled together a remarkable planning group, including their astute team leader Len Ruggiero. As with any organization, however, office politics came into play. Jack Booth had a poor relationship with his boss, Craig Rupp, the regional forester in Denver. Rupp did not trust Booth to develop a credible plan, but Rupp did respect Ruggiero and his planning team. The staff officers and rangers remained pretty clueless about how the planners did their business, but they adopted a "Guess we better trust 'em, because we sure as hell can't tell 'em how to do it" attitude. We counted on Ruggiero to lead us.

Regional foresters pushed hard to ensure that each national forest did its part to contribute timber harvest to the big pot for the entire agency. Money flowed from Congress with a high expectation that the Forest Service would sell a certain amount of timber for a certain amount of money spent. If you met your timber target, you were a "good man," a leader who was trustworthy, respected, and made of the right stuff. If you failed to meet your timber target, well...another job better suited to you awaited. Bear in mind that at this juncture, about 1980, newly elected President Reagan had appointed John Crowell, a timber-industry lawyer from Louisiana-Pacific Corporation, as under secretary of agriculture to oversee the Forest Service. Crowell's memoranda to the Forest Service spoke of his belief that the timber cut from national forests could and should be doubled. Such timber harvest levels were unprecedented.

The Bighorn forest supported Wyoming Sawmills, a monopoly sawmill in Sheridan. This mill and the Forest Service engaged in an ongoing battle over how much timber would be sold and at what prices. With good lumber markets, Wyoming Sawmills couldn't get enough wood. Conversely, bad markets made them reluctant to purchase logging contracts. Wyoming Sawmills and the Bighorn had a fractious and rancorous relationship. But, in truth, we needed each other if there was to be any logging.

Inevitably, Regional Forester Rupp and Wyoming Sawmills took a vital interest in Bighorn FORPLAN predictions. So did the district rangers because we'd ultimately have to deliver the goods. Historically, the Bighorn produced about seven to ten million board feet annually.

As I recall, the first modeling projections showed estimated yields of about twelve to thirteen million board feet. We rangers unanimously protested these numbers as "silly science." The computer might have generated that number, but we viewed it as unattainable in the real world, based on our practical experience and sensibilities. Regrettably, we had little factual data with which to rebut the FORPLAN model.

How reliable is gut instinct? The guts of all five of us rangers said that decades of experience ought to count for something. I don't recall explicit discussions about our perceptions of the consequences of chronic overharvesting. I think all of us weighed timber harvest against elk and trout, beautiful creeks, pristine recreation in roadless country, solitude easily found, the "feel" of a country. We knew much of this would be lost pursuing timber worth but a pittance compared to the values at risk.

Rupp, meanwhile, demanded that those nasty little assumptions be tweaked to see what the Bighorn could really produce. The Bighorn's planners "adjusted" the FORPLAN model and somehow muscled out about twenty million board feet. All things considered, not a large quantity of wood, but it more than doubled the historic harvest level.

For the first time in my career, I felt a queasy discomfort about the consequences of decisions in which I would be complicit. I sensed that higher powers would roll us, trading away environmental principles for economic values.

Could this be the Forest Service I loved, respected, and admired?

I become deeply attached to a piece of country when working there. I find it hard not to. I don't know quite how to describe it, but land speaks its innate character into my soul, and a bond grows. I confess to having a greater affection for some places than others. For example, I care for the Bighorns more than the forests of Maine. I suspect most people develop some kind of kinship with land. I know this phenomenon affects people and powerfully shapes culture. Farmers and mountain folks both relate strongly to land and nature but have evolved very different cultures.

The Forest Service inevitably evolved a culture over time that reflected its working environment and an ethos about its land management mission. The ethos of the agency, one of being a steward of public lands, should change over time to accurately reflect shifting public values. I readily confess that the voices of townspeople, ranchers, and other Forest Service associates I heard in Wyoming were pretty comfortable with the status quo, and that made it easy for me to ignore or not listen for other dissonant voices. But those voices were out there, as I discovered when I later started reading *High Country News* and other "radical" material.

About 1978, I stood next to a small creek in the vicinity of Soldier Creek with Tensleep District staff and several specialists, including a soil scientist, wildlife biologist, and hydrologist, discussing steps needed to close a timber sale. This one happened to be the oldest agency sale still operating in the entire country. I stated my preference to keep a buffer of uncut trees alongside this creek. Bob Damson, the forest silviculturist, challenged this and asked me why. I said it would protect the creek from possible logging damage. Even forty years later, I still smart when I recall the sneer on his face when he said, "What are you doin' working for the Forest Service?"

Damson and I had a different land ethic. He felt a buffer was unnecessary and wasted good timber. I disagreed. Yes, I thought my land ethic better than Damson's in this instance. In fact, the Forest Service now requires such buffers everywhere. The agency's ethos eventually changed to reflect a greater respect for water quality and protection, as it should.

Damson's land ethic implied that good forest management necessarily involves tree cutting and wood product yield as a dominant value. My ethic allows cutting trees only after seeing to basic environmental safeguards. But Damson's ethic carried the day in guiding how much timber to harvest on the Bighorn. We published a forest plan with a timber harvest level of seventeen million board feet. We rangers understood the exercise of power. Regional Forester Rupp got his timber.

Senior leaders in the Forest Service of the 1980s had been trained by dedicated, hard-charging men of strong convictions who created and adhered to a timber-first dogma. Future leaders in my age cohort had serious reservations about the wisdom of the elders.

Experience

"Experience?" My district forester, Tony Aaron, used to laugh, saying, "Experience is what you get when you thought you were gonna get something else."

I was getting some experience. As one year in Wyoming became two, then three and four, the luster of my dream job began to wear thin. I came to the Tensleep job as the youngest ranger in the Rocky Mountain Region. I was quite proud of this distinction, though it lasted only a year. A younger Denny Bschor, an Iowa State classmate of mine, settled in at Paintrock District just north of Tensleep. Before arriving at Tensleep, I'd convinced myself that I was damn good and ready for a ranger job and deserved my shot. This notion proved wrong.

I learned years later that Jack Booth, the forest supervisor and my immediate boss, had not chosen me. The regional office had placed me, and I confronted a very skeptical Bighorn organization that offered little help and support. The responsibilities of the ranger job had overwhelmed me and exposed my inexperience. I'd learned that the previous ranger had left under a cloud, leaving behind a disorganized staff and poor morale. Things certainly couldn't get much worse. It didn't seem I could drag my unit much further down.

After I'd been at Tensleep about a year, Booth shocked me by saying, "You're wearing yourself out trying to get the best out of the staff." He bluntly told me they wouldn't come around. I needed new blood. Perhaps an organization's surface appearance, just like the appearance of the land, can deceive. I tended to think I needed to make do with the hand I'd been dealt. Booth asked me to look deeper and I'd see my staff was not healthy. I needed to manage my business—make changes or Tensleep District would continue to founder.

I reluctantly agreed, saying I needed him to help me find other jobs for my folks as well as to locate good, skilled replacements. He agreed. I got no help whatsoever.

Yet changes did come in the next couple of years. Booth was right—we needed new blood. When it came, good things happened. I worked hard to find transfers for some of my staff, and, when openings came, I looked hard for the best people I could find. I shared blunt truths about circumstances on the district—it was a mess and I needed patient builders. But the country was spectacular, and Worland was a fine town for raising a family. I learned an important principle: if your staff can't or won't change and grow to work with you, find people who will. I would need to apply lessons learned at Tensleep again.

Safety became a big priority. The Rocky Mountain region had the worst safety record in the entire Forest Service, the Bighorn had the worst record in the region, and, of course, Tensleep had the worst record on the Bighorn—rock bottom. At one point, Don Ackermann, a forestry technician and a top hand, came to me, embarrassed, to say they were running a snow machine with a missing front ski and didn't have helmets to wear. He was afraid to ask me for the money to correct this for fear I'd tell him to stop complaining and get back to work.

It was an easy call to get the ski fixed and buy helmets. But how could we have sunk so low?

With much hard work, we eventually turned our safety program around and actually came to have a good record. I learned a good deal about safety management in the process. I had to swallow the fact that my managerial deficiencies contributed directly to our poor safety record and put my staff at risk of personal injury.

I learned another hard truth: first impressions last a long time. I made my share of mistakes and blunders and failed to earn the trust of my peers. Though Jack Booth and I got along well personally, we often didn't see eye-to-eye on professional issues. I received a letter of reprimand after ordering some rock for an erosion project on Webb Creek without getting prior authorization for the procurement from staff in Sheridan. I thought extenuating circumstances from my back surgery should have exonerated me, but technically I was in the wrong—my lapse showed inexperience and incompetence. Some people thought me lazy because I didn't make a habit of going to work at the office on weekends and even told me I'd "never make it in the outfit" if I didn't buckle down.

But I knew full well the pivotal event that really ripped it with Booth. The Canyon Creek timber sale needed a final inspection before we could close the contract. I met Dick Weed, our engineer, and Ernie Schmidt, the head forester with Wyoming Sawmills, on Highway 16, where we discussed our objectives briefly and then headed down Canyon Creek Road. Final road maintenance stood out as a critical concern because the regional office in Denver had just issued a scathing road review of the Bighorn. Booth made a point of telling every district ranger that all timber roads should be built to specification and maintained as such. No exceptions!

A couple inches of fresh snow gave the woods a pristine, sparkling look, and our boots squeaked as we headed up the first road, drawings in hand. I checked to make sure the road surface was graded to remove all ruts and that water bars to drain runoff were in place to handle spring snowmelt. Everything looked great, except the water bars were in the wrong place. After noting that the first three water bars were not installed as per the road plans, I took Dick Weed aside and asked for an explanation.

Weed said that he'd used the plans as a "guide" during road construction (about four years ago) but had made adjustments on the fly. So the road on the ground didn't match the road on the plans. The changes were recorded and approved "as built" in his original inspection diaries. I asked Weed if I could expect to see this elsewhere. "Yup— pretty much all day long."

I felt like a trapped coyote, sensing steel jaws clamp shut on my leg. My mind raced. Ernie Schmidt had gone to a contract disputes board several times to overturn Forest Service decisions. I knew if we didn't accept his road crew's final maintenance work to polish all the roads, he would dispute our decision because he'd built the roads as agreed to by the Forest Service and done a nice maintenance job.

On the other hand, Jack Booth would be outraged if I acted in direct violation of his order, and I could anticipate serious repercussions. It seemed the choice was either give Schmidt the closure he deserved or tell him to come back next summer to rebuild all the roads to the original specifications. In that case, Wyoming Sawmills would soak us for some serious damages.

We looked at all the roads, and I informed Schmidt that the sale would be closed. I went home to compose the official closure letter to Wyoming Sawmills in my capacity as contract officer and prepared for my dreaded phone call to Booth.

It hit the fan as expected. Because I had been delegated authority, Booth could not reverse my decision; it was final. But he was livid. He saw my action as insubordinate, gutless, and unprofessional. My reputation with Booth suffered irreparable harm. I thought my decision was simply common sense and saved the Forest Service acute embarrassment and a lot of money.

Even though I felt I continued to learn and grow to become a competent ranger after a few years, I was wrong to simply assume I'd eventually move up the ladder to a more prominent ranger job.

About 1983, Jack Booth retired early because of his persistent conflicts with Craig Rupp, drummed out of the agency he dearly loved. Sadly, he died of a virulent intestinal cancer only a few years later.

Ed Schultz replaced Jack Booth to become the next Bighorn National Forest supervisor, with instant impact. Schultz was firmly in charge from day one, and I witnessed a hands-on, focused, determined manager at work. I could feel Schultz sizing me up at all times. But then, Schultz sized everybody up at all times. It was kind of fun watching him operate. A bit scary, too.

Within a year, Schultz and I had "the talk." October rolled around, and with it, time for my annual rating. As we enjoyed a beautiful October day cruising around Tensleep District, Schultz eased into asking how I saw my future unfolding. I laid out my plan of moving on to a bigger, better ranger job. Schultz slowly lit his pipe—long pull, exhale—then he simply said, "Not gonna happen."

He went on to explain, matter-of-factly and with some compassion, how my reputation was in the toilet. Other forest supervisors were unwilling to take me because Booth had poisoned the well. I had a reputation, and not a good one. I was considered too risky to take a chance on promoting to a bigger challenge. Schultz said he'd evaluated me carefully and actually thought I was OK. Maybe I'd gotten a raw deal, but the truth mattered, even if unfair, and I better deal with it. Either that or be prepared to sit a long time.

Such blunt truth has a way of clarifying things. I held onto that in case I needed to use it myself some day. Eschew obfuscation.

One thing I admired about Ed Schultz: he made things happen. Not a month passed before he asked me to consider two jobs, both forest planner positions in southern Colorado, one on the Rio Grande National Forest, the other on the San Juan.

I went home to share news of these opportunities with Judy. We discussed the apparent inevitability of a move. Her parents had already made plans to take their three daughters and sons-in-law to Hawaii, and we wanted assurances that these vacation plans would be honored. The

San Juan National Forest had its offices in Durango, Colorado, which was regarded as a great town. The forest itself had a sterling reputation as a great place to work. The Rio Grande National Forest, in Alamosa, Colorado, compared poorly to Durango in almost every way.

I told Schultz I would accept either job, but I much preferred Durango. He said, "Good answer." Within two weeks I received an offer to transfer to Durango. God had smiled on me again, as I was to discover in the ensuing years.

We had the usual going-away party and said awkward good-byes. Many friends we'd never see again. We had a house to sell, a house to buy, and a new job to adjust to. When you work for the Forest Service, you learn how to do this.

The Tensleep Ranger District is no more. It merged with the Buffalo District to form the current Powder River District, headquartered in Buffalo. I have enjoyed returning to this great country many times, most recently in 2011 on our road trip with my two brothers and two sisters. I thoroughly enjoyed the luxurious pleasure of sharing so many good memories with my siblings.

As I write this, the Powder River district ranger is none other than Mark Booth, second son of Jack. He, like me, became a ranger in the mountains of his childhood. Mark Booth is a fine man and a real professional. The old Tensleep country looks even better than when I left in 1984.

I eagerly anticipated a fresh start in Durango, but I felt like damaged goods. I'd arrived at Tensleep with stars in my eyes and left with a shiner. As the weight of responsibility settled on my shoulders, I came to sense I hadn't made the grade as a district ranger, certainly not in the eyes of many of my peers, nor, to a degree, in my own eyes. I left with great memories of wonderful country and wonderful people, but also with a bitter awareness that maybe I wasn't quite up to the challenges I faced. Durango was do or die.

Chapter 4: Redemption and Revelation

SAN JUAN NATIONAL FOREST, 1984–1989

Past ten now, time to chunk a couple of pieces of piñon pine into the wood burner and head for bed, thankful for flannel sheets. I expected the usual nice bed of coals in the morning, enough to get a new blaze going with fresh wood and a wide-open damper. A quick peek at the night sky revealed a brilliant, clear, moonlit sky. I awoke the following morning to another bluebird day with a radiant Colorado sun coming up over the ridge east of Durango.

To my surprise, a fresh six-inch cap of snow sat atop the deck railing. Clear sky last night, clear sky now, so where did this snow come from so stealthily in the night? After living in Durango for five years, I repeatedly saw snow magically come straight down, heavily at times, to pile up nicely on any surface—car, mailbox, fencepost, railing, tree limb. Only once did I witness snow and wind simultaneously.

Southwest Colorado had a startling beauty and purity to it on mornings like that—cold, but never bitter, with enough sun to make it comfortable to be outside in a heavy shirt and vest. Rimmed by the La Plata Mountains west of town and the distant Needle Mountains to the north, Durango wintered on while the Animas River coursed quietly through town. I could understand why people loved it here.

We'd arrived in February and departed shortly for our long-anticipated trip to Hawaii with the rest of Judy's family, entirely paid for by her folks. What a great gift! I remember the trip fondly, especially body surfing at Makapu'u on perfectly curling ten- to twelve-foot waves. Colorado's winter was a vivid, beautiful contrast to Hawaii's warm ocean.

Now I needed to face the music at work. I felt like damaged goods, very insecure and seriously wondering whether I had a future with the Forest Service. I speculated about what rumors had accompanied me here.

Much to my relief, John Kirkpatrick, the forest supervisor, welcomed me warmly and expressed sincere enthusiasm about my joining his staff. Kirkpatrick also was new, having arrived shortly before I did. If he faked his welcome, it was a good act.

I quickly got acquainted with folks in the Durango office, then attended one of the monthly leadership meetings of staff officers and district rangers. The meeting was an open, free-flowing, energetic dialogue, with everyone eagerly participating. I couldn't help but think how different things were here than on the Bighorn.

There, I recalled, everybody lay low all day, volunteering little in the way of opinion or information. The goal seemed to be to risk nothing, just get through the day. Then the rangers (only the rangers) would head to the bar for beer and popcorn. Then the real meeting began. Beer loosened tongues as we rangers began to parse the day, discussing our schemes to accomplish what we wanted and how to block what we didn't want. Part of the drill included my fellow rangers reminding me that I needed to learn to keep my mouth shut during the day's meeting. I never felt comfortable about this gamesmanship. The new team on the San Juan seemed refreshing and satisfying. Had the Bighorn really been that dysfunctional? Yes, I think so.

Planning Redux

The San Juan's forest plan had been completed. I would be the staff officer responsible to implement the plan and monitor its performance. As an ex-district ranger, I felt more comfortable in this role than the highly technical role of preparing a plan.

But not so fast! What seemed like a straightforward job turned out to have a twist. The San Juan forest plan had been appealed by several environmental groups, requiring a review by the chief's office. The chief remanded the plan due to deficiencies in the analysis of the timber program, necessitating a comprehensive revision of the plan and a new Environmental Impact Statement (EIS). A pattern quickly emerged all across the country—every forest plan was being appealed.

The National Forest Management Act created a forest planning process intended to identify and solve problems by addressing competing interests and judiciously balancing programs for all resources. At least, that was the idea. In practice, forest plans served as an anvil where sparks flew to ignite intractable disputes rather than being a crucible for credible compromise. In large part, forest plans held to the agency's timber-first priority. Environmental groups waged war. The chief's office repeatedly remanded plans back to forests to address and resolve deficiencies, as in the San Juan case, and to strengthen the rationale for our timber program. If litigation followed, we needed to have as strong a case as possible.

The San Juan forests differed significantly in character from those of the Bighorn. This country had beautiful, expansive ponderosa pine forests, some of the biggest and best aspen in the world, and higher-elevation spruce and fir. Timberline ran up to 11,500 feet this far south, much higher than on the Bighorn.

These factors, coupled with the fact that the San Juan forest acreage was twice that of the Bighorn, resulted in a projected annual timber harvest of forty-one million board feet (mmbf) or allowable sale quantity (ASQ), compared to seventeen million board feet on the Bighorn. The FORPLAN model, which was widely used by the Forest Service to plan timber harvest, indicated that the forty-one million board feet ASQ was biologically feasible and sustainable.

An important part of the equation was related to timber industry capacity in the area. The history of San Juan timber harvesting followed a nearly identical pattern as on so many western forests. In southwest Colorado, a small custodial logging program existed until the mid-1950s, followed by a rapid escalation in logging after the installation of two big pine sawmills, one in Durango, the other in Pagosa

Springs. Loggers punched roads into far corners of the forest to bring timber to town. As the annual harvest often exceeded one hundred million board feet, the timber industry blossomed into the economic mainstay of many local communities.

Then, after a good twenty-five-year run, many logging companies fell on hard times during the recession in the early 1980s. Durango's San Juan Lumber Company was in its final death throes when I arrived in 1984, a reality that made it difficult to assess future industry demand. Furthermore, a new buzz ran through Forest Service offices—after the easy pickings had been logged, timber was getting hard to find.

The San Juan forest plan was an acknowledgment that the big timber days were over, but even the new level of forty-one million board feet would be difficult to provide. Environmental groups, led by Colorado Environmental Coalition, appealed the forest plan, challenging forty-one million board feet as being unreasonable, unsupportable, and unsustainable. The timber industry, for its part, sought to protect its vital interests.

I would lead a review of the plan's analysis and conclusions, try to equitably balance competing environmental and industry interests, and amend the San Juan plan accordingly. We had the freedom to provide more credible support for the plan's forty-one million board feet timber harvest level or adjust to a new ASQ level, most likely lower. We needed to involve timber industry and environmental groups every step of the way and finish our work within two years at most, at as little cost as possible. I needed a secret weapon. I knew one.

He was Jim Powers, a very sharp forest economist who'd worked as an analyst for the Bighorn forest planning team. The planning effort was now concluded on the Bighorn, which meant that Jim and other Bighorn planners would be looking for work elsewhere. All of them had unique experience and skills, but Powers was reputed to be the analytic engine for the Bighorn because of his good working knowledge of FORPLAN. This acumen would be essential in preparing our plan amendment. I regarded Powers as nearly a genius, based on my observations of his analytical skills.

Getting Powers to work on our plan amendment would be a coup. After discussing options with John Kirkpatrick, I approached Powers about a two-year temporary assignment in Durango. We sealed the deal.

Two primary protagonists participated from start to finish. Rocky Smith represented the Colorado Environmental Coalition (CEC), and he was well-known by all my peers on other national forests as an intense, passionate warrior for the environment. Frank Gladics represented the Intermountain Forest Industry Association (IFIA). Gladics, a big bear of a man, had worked for the Forest Service before his move to the timber industry, so he understood how to move the levers of power within the Forest Service.

We also needed to heed the desires of other "masters of destiny," such as our San Juan leadership team, the regional forester, and lawyers with the Office of General Counsel (who served the flagship USDA). Each party effectively had veto power if its representatives could not be persuaded as to the efficacy of our solution.

Our work would be precedent-setting, since the San Juan forest plan was the second in the entire nation, and our revision would be seen as an example for others to follow. Litigation loomed as an inevitable outcome, if not for the San Juan plan, then elsewhere, since opponents would be probing for weaknesses to preclude or reduce logging on all national forests.

Many foresters on the San Juan doubted that a harvest level of forty-one million board feet was sustainable. They had misgivings like those that registered on the Bighorn. We now had an opportunity to get it right. Jim Torrence replaced a retiring Craig Rupp as regional forester. Torrence came to Denver from his position as deputy regional forester in the Pacific Northwest. Rumor had it that he was not as autocratic as Rupp, and I hoped he would be more reasonable and flexible on timber harvest levels.

The FORPLAN timber harvest model had a reputation as a naughty child. Complicated and difficult to manage, the computer model could be strong-willed, producing outcomes that left you wondering, "How did that happen?" Norm Johnson had developed FORPLAN to model timber production, but it had been modified by the Forest Service to estimate other resources as well, while tracking costs and benefits, so that it served as a comprehensive planning tool.

I was aware that FORPLAN fostered a timber-centric planning process, even as public sentiment was questioning the environmental sensitivity of the Forest Service. But a more urgent problem for our team now was that FORPLAN crunched its numbers impersonally, working tirelessly to optimize timber production. It did little to identify mistakes, let alone fix them.

To address this, we had the invaluable expertise of Jim Merzenich, who worked for the Forest Service in Portland, Oregon, and who volunteered to work with us. Merzenich, a gnarly ex-wrestler with telltale cauliflower ears, didn't seem to me to fit the stereotype of a computer geek. But he created a number of editing routines capable of critically analyzing a FORPLAN solution, arraying the information in more easily understood formats.

One Merzenich edit routine, for example, displayed all the timber stands FORPLAN selected to be logged, arrayed from the most volume to the least. We could readily see acres, volume of timber per acre, tree species, and so on. This type of analysis proved crucial in helping us evaluate the FORPLAN solution that had been the basis for the San Juan forest plan's timber level, which had provoked the appeal.

Powers, Merzenich, and I eagerly awaited the printout churning on the printer. When we took a look at the numbers, Merzenich moaned. I thought I might get ill.

Perched on top of the list, quite proudly, was a three-thousand-acre stand of spruce listed as having a volume of about three hundred thousand board feet per acre. This constituted a mistake of enormous magnitude. The figure was about a hundred times bigger than it should have been—normal volumes were only about three thousand board feet per acre. The FORPLAN model loved this stand and harvested it repeatedly over time.

This simple, small error had enormous implications. It meant that this one small stand yielded ten million board feet out of the forty-one million total harvest.

This discovery left me first shocked, then scared at the thought of trying to explain it. I acknowledged that the mistake, if it came to light, could cast doubt on the integrity of the entire forest plan. On the other hand, once corrected, it would immediately drop the harvest level to a much more reasonable and achievable level. Gladics and IFIA would be outraged that ten million board feet a year could disappear due to a misplaced decimal point. Smith and CEC would be exultant, but suspicious of other yet-undisclosed errors.

We now had an issue of integrity—whether, and how, to disclose this glaring error.

After some deliberation, I elected to put a close hold on the mistake. I certainly didn't intend to cover up the gaffe, but our analytical process had just begun, with much more work required before we could even attempt putting all the pieces together to reach a consensus agreement. Other factors would likely affect our process to develop a more reasonable timber harvest level. We plowed on.

Another important procedure involved evaluating economic elements of cost and benefit. The Forest Service has a long history of selling timber to logging companies. Jim Powers undertook a meticulous analysis of San Juan timber sales. He sought to develop, if possible, a downward-sloping demand curve that accurately represented the San Juan's economic environment for timber commodities.

Simply put, one expects that demand for timber will decline as prices rise higher and higher. Conversely, as prices decline, one would expect demand to increase to capture more profit. This elastic relationship should be reflected in real-world financial transactions for timber, just as it is for oil or beef or corn. Powers told me that no national forest had ever analyzed timber demand in this way for its local market, even though required to by agency planning policy. Most forests lacked the technical staff capability to do this credibly.

Powers's rigorous analysis yielded a demand curve based on a sophisticated mathematical regression analysis and testing of secondary measures (my personal favorite: *heteroskedasticity*). The demand curve provided a rational economic basis to suggest a narrow spectrum of about twenty-two to twenty-six million board feet of annual timber volume to offer for sale.

Put more simply, even if the land could grow a greater amount of wood, the San Juan shouldn't sell more timber than the market demanded. This constituted a meaningful alternative to the timber harvest regime predicted by FORPLAN.

We now knew that, minus the ten-million-board-foot mistake, the optimal FORPLAN level of sustainable harvest was no more than thirty-one million board feet. And we had an economic, market-based argument for no more than twenty-two to twenty-six million board feet.

A solution seemed suddenly closer. The chief had asked for a credible forest plan, Smith and CEC wanted a reduced timber harvest, Gladics and IFIA had to admit that previous mistakes eroded support for the original forty-one million board foot harvest level—and this result would likely stand up in court. I felt confident that we had done our job well.

Hard work remained, but in ensuing meetings we convinced Gladics and IFIA that a twenty-four million board foot harvest level could adequately supply and sustain the local timber industry. We also persuaded Smith and CEC that this allowable sale quantity had enough slack in it to avoid pressing hard against the sustainable capacity of the land.

When we explained our analysis to the San Juan's leadership group and staff foresters, they expressed that, intuitively, this lower ASQ seemed a good fit. Regional Forester Torrence expressed two concerns: first, whether our analysis could withstand legal challenge, and second, whether the outcome enjoyed sufficient support from the opposing parties to dissuade them from further appeal and litigation. We gave assurances that we stood solid on all counts.

Thus, a bargain was struck for twenty-four million board feet, a timber harvest level substantially less than the land could produce sustainably and sufficient to provide for local timber mills.

We completed all necessary environmental documents and issued the decision. We had no further appeals. All in all, the process succeeded on many levels. We completed the job on time, and at relatively little cost, especially compared to other, similar plan-revision efforts. Virtually all parties came away feeling understood and with an outcome that addressed their interests. The agency benefited by having a workable model to build on, since virtually every national forest would have its timber program challenged. We'd laid down a good precedent.

Working through this challenge over a period of two years went a long way toward restoring my confidence. What seemed like an intractable problem had, in fact, yielded to a good outcome. People behaved like reasonable adults. My feelings of being a district ranger nobody wanted faded. John Kirkpatrick's trust in me had grown. John now had a new challenge for me to work on.

Twin Peaks

Heading east from Pagosa Springs toward Wolf Creek Pass, one passes Wolf Creek Valley, a gorgeous sprawling ranch astride the San Juan River. Harvey Doerring, a wealthy real estate investor, purchased Wolf Creek Valley in 1985 and approached the Forest Service about building a large new ski area on adjacent national forest lands. Doerring planned to develop this sizable private land parcel with the typical pattern of ski lodges, condominiums, and homes. There had been no significant new ski area built in Colorado for many years, and the controversy surrounding Doerring's proposal was enormous and instantaneous.

The development of America's tremendous ski areas following the Second World War owes largely to veterans of the Tenth Mountain Division, who understood that terrain rivaling Europe's great ski resorts in the Alps lay almost entirely on national forest land. Pete Seibert and a handful of ambitious entrepreneurs, after climbing to the top of Vail Mountain in 1957, approached the Forest Service to inquire if national forest lands could be used for downhill skiing. The agency said yes in the belief that ski-area development was in the public interest. Encouraged by the positive response, Seibert secured financing to purchase private land and develop a ski resort, and Vail opened in 1962. Many major ski resorts now operate under Forest Service permits using the principle of private money expended on public land to create recreation opportunity. This partnership between the ski industry and the Forest Service spawned explosive growth in ski development from 1960 to 1980.

The decision whether to approve the proposed Wolf Creek Valley ski area sat squarely in San Juan Forest Supervisor Kirkpatrick's lap, since he had authority over any private business proposal on public lands. Transformation of the bucolic valley bottom hung in the balance. If the Forest Service approved a new ski area, the historic ranch property would almost certainly be transformed into a densely developed ski town. If the Forest Service said no, Doerring's speculative gamble on acquiring the land would likely result in eventual resale of the property, possibly at great personal financial loss.

Doerring had reason to be optimistic—Vail, Aspen, Breckenridge, Winter Park, Steamboat, and other ski areas had been approved. But that was then; this was now. Strong opposition toward ski-area development had emerged as part of the environmental movement, particularly toward new areas, with expansion of existing areas preferred.

Because this project would require a massive environmental impact statement, Kirkpatrick pulled me in to help coordinate Forest Service activities. Bob Lillie, our recreation staff officer, also played a key role. We three conferred and agreed that we needed one person as project manager. Duties would include managing community relations with Pagosa Springs, about fifteen miles west of Wolf Creek Valley, as well as coordination with state and county governments. We needed to establish standards for Doerring's consultant (who would do the environmental analysis) and then monitor the progress. All Forest Service involvement would need tight coordination to avoid duplication. We selected Don Hoffheins, a soil scientist with a gift for project management, to handle Wolf Creek Valley.

As this project got underway, we were surprised when another ski area proposal emerged. Dan McCarthy, a wealthy businessman from Evanston, Illinois, owned a twelve-hundred-acre property on the East Fork of the San Juan River, a few miles east of Doerring's holdings. McCarthy's East Fork ski area was to be even larger than Wolf Creek Valley. These two proposals, in tandem with the existing Wolf Creek Pass ski area, would transform Pagosa Springs into a new epicenter of destination skiing.

What prompted McCarthy's timing? He had owned East Fork for many years, buying it after methodically searching the Rockies to find the perfect place for a megaresort. I sensed that he was provoked to act because he perceived Wolf Creek Valley as a threat to his dream and put forward his proposal to create a "survival of the fittest" showdown. It's also possible, though less likely, that he anticipated a synergistic effect, with both ski areas benefiting from the presence of the other. I do know that McCarthy felt East Fork to be vastly superior to Wolf Creek Valley. He told me so many times! Importantly, McCarthy held to a belief that if the Forest Service were to approve only one ski area, it must be East Fork.

Both sites had outstanding potential, each with spectacular scenery, abundant snowfall, a good mix of skiing terrain for all abilities, and ample vertical feet from valley bottom to mountain top (over three thousand feet at East Fork) to be a premier resort. Wolf Creek Valley had the advantage of being along US Highway 160. East Fork access twisted through a narrow canyon that would require extensive improvements. But the consensus among Forest Service folks like me was that East Fork was the most magnificent ski mountain.

There were important environmental issues at both sites related to water quality, elk habitat, scenery, and effects on nearby wilderness areas, as well as persistent rumors of grizzly bear in this vicinity. These environmental impacts could not be eliminated, but we felt they could be minimized to a reasonable degree.

However, the crucial issues would be social and economic. These impacts would be huge, unavoidable, and viewed as either positive or negative depending on one's perspective. If even one of the ski areas were to be built, we knew it would completely transform Pagosa Springs. Some people desperately wanted an amped-up economy—new jobs, construction, improved highways and airport, upgraded water and sewer systems, burgeoning business growth—and the prospect of name recognition for tourism. Many others felt the town they loved would be lost forever. How can such things be measured?

Other concerns involved increased crime and drugs, seasonal shrink-swell of low-wage jobs, the probability that the real money would gravitate to the ski areas and not to the town of Pagosa Springs. There were complaints about the ubiquitous ski bums who, it was assumed, wouldn't care about the community. Pagosa didn't want thousands of "takers." The concerns expressed at town meetings, in newspaper opinion pieces, and by local government officials were numerous and valid.

City and county elected officials tend to be pro-business and pro-growth, especially in lightly populated areas like Pagosa Springs. As expected, they supported ski-area development, but we had to consider other views as well. Pagosa has a long and strong history as a Latino community, and people feared losing this culture. Many people moved to Pagosa for its small-town feel. A ski town threatened their values.

As we deliberated over these social concerns with citizens and government officials, the two developers, Doerring and McCarthy, faced financial hurdles. Almost all growth in the ski-area market had been accomplished by upgrading and expanding skiing at existing areas, such as Vail, Aspen, Winter Park, and Breckenridge. Environmental and regulatory reviews were rigorous and costly. For investors interested in ski development, expansion seemed more prudent and less risky than new ventures. Even wealthy individuals like Doerring and McCarthy lacked the personal wealth to wheel and deal a ski-area development without securing money from outside investors.

Review of ski area proposals by the Forest Service revealed that financing was something of a chicken-and-egg game. How does a ski-area developer attract investment capital when the ultimate question of Forest Service approval remains unanswered? That's because the Forest Service would rather not extend a firm decision until the proponent can demonstrate secure financing in hand to finish the project.

Because we Forest Service officials weren't bankers, evaluating proponent financing required that we look at secondary factors such as whether the applicant held valid title to the private lands that would serve as the base of the ski area; whether he could afford to hire a private firm to conduct all the environmental reviews, usually at a cost of several hundred thousand dollars; and whether his personal financial statement indicated sufficient resources to realistically pursue a major project. We did our best to determine the financial integrity of both Doerring and McCarthy. We deemed both of sufficiently sound financial health to proceed.

Doerring's Wolf Creek Valley got the jump on McCarthy's East Fork. Doerring accomplished all the work and analysis necessary for the Forest Service to complete this Environmental Impact Statement first. The lengthy Wolf Creek Valley EIS provided an assessment of options and their environmental effects.

John Kirkpatrick met with his principal staff throughout the process to achieve consensus in describing key decision criteria for the proposal. Kirkpatrick assigned me the task of fleshing out a draft record of decision (ROD). An ROD is the comprehensive, coherent narrative that articulates the criteria for a decision and the rationale behind its conclusions.

I recall holing up in my office for about a week after Kirkpatrick gave me the assignment. I churned out about a twenty-pager, which then got a rigorous internal review and edit. My draft ROD addressed issues like elk winter range, water quality, jobs and economic impacts to Pagosa Springs, and, crucially, whether estimates of skiing demand warranted development of a large new ski area.

After weighing all relevant factors, Kirkpatrick approved the Wolf Creek Valley ski area. We specified that a permit to develop the site must be issued within two years or the decision would be voided.

Then, with the entire EIS process completed, Doerring declared bankruptcy! We were stunned. Doerring had given us no indication that he was in financial peril. Wolf Creek Valley slowly and finally sank into a financial quagmire. The valley today looks much as it did in 1985.

East Fork, fueled by McCarthy's optimism, remained a potent proposal. McCarthy, a bit smug about Doerring's collapse, moved forward at full speed. The process unfolded similar to that for Wolf Creek Valley, although more efficiently.

In the end, a decision on East Fork turned again on familiar factors. Did estimated skiing demand support building a large new area in southern Colorado? Did anticipated socioeconomic benefits override negative environmental and local community consequences? Did elected officials support the project? Environmental groups focused primarily on degraded water quality, widening and paving the road up the canyon, and developing a pristine valley.

I recall many spirited discussions on the question of demand. The Forest Service occupied a unique position with respect to skiing in the United States, since most of the best ski terrain in the country is on national forest land. The explosive growth of downhill skiing could not have happened without the unique partnership between the Forest Service and the ski industry.

As with logging and grazing, the Forest Service strives to make reasonable accommodation for free enterprise while protecting environmental values. The Forest Service usually said yes to proposals to expand capacity at ski areas that asked for new lifts and runs. But few proposals had been received for new areas in recent years.

The real debate on public land skiing by the late 1980s was whether all growth in supply should be confined to and accommodated by expansion of existing areas. Or was there room at the table for new players?

A free-market position held that for a healthy, resilient ski industry, the Forest Service should not artificially constrain the market. Proposals for new areas should be judiciously evaluated, with the best prospects given an opportunity to enter and influence the market. If some aging, lower-quality ski areas folded, so be it. No industry could long survive if not given the opportunity to change and grow.

Could a new site attract investment capital? A certain truth played out in the market: capital typically flowed to ski-area expansions. A loftier threshold existed for financing new areas, largely based on their greater risk and uncertainty.

The analysis for East Fork supported a finding that it had outstanding qualities as a major ski area. McCarthy exhibited a strong commitment to minimizing environmental consequences, even undertaking a costly restoration of the East Fork San Juan River on his private land. Elected officials were supportive. Demand projections supported the presumption that a major high-quality destination ski area could muscle its way into a competitive market. In our collective mind, the sum of all factors favored authorizing the ski-area proposal.

Once again, we issued a yes decision, approving East Fork with conditions, including the proviso that financing be secured as well as a special-use permit issued by the Forest Service within about two years, or the approval would be voided.

Ultimately, McCarthy could not secure additional financing sufficient to proceed, though he certainly tried. East Fork did not undergo bankruptcy like Wolf Creek Valley. It languished and remains today much as it was, a spectacular secluded valley surrounded by some of the best undeveloped skiing terrain in the United States.

I supported both of these proposals. I confess I enjoyed downhill skiing a lot (though I've retired from the sport), and a part of me wanted to see this mountain thrill thousands of people. Another part of me is relieved that the East Fork ski area never happened. Even though the Forest Service approved the proposal, larger forces beyond our control prevented East Fork, and there is a certain justice in that outcome. But I still wonder what will become of Wolf Creek Valley and East Fork.

New Possibilities

I found working on challenging issues on the San Juan enjoyable and professionally rewarding. John Kirkpatrick and I forged a very close professional relationship. I came to respect and admire him as we worked together on difficult issues. Our trust in each other was mutual and deep. What a pleasant outcome given how deflated and disenchanted I had been a few years earlier. Kirkpatrick shared with me just about everything that was going on.

Thus, I was more shocked than almost anyone when John announced he'd accepted a position in the chief's office. I hadn't even known he'd applied! Reflecting on this, I realized a boss needs to keep a close hold on some things. He was being smart, not insensitive.

We soon received word that our new forest supervisor had been selected—Bill Sexton was coming to Durango. It seemed no one knew much of anything about him. Rumors filled the void. We knew he was coming from the chief's office, so it was expected that he would arrive with a big-picture perspective. Some people had heard Sexton was an American Indian, and that his being a minority clinched his selection. It was said that he was an up-and-comer on a fast track for advancement.

Sexton and I hit it off immediately. He worked really hard to meet a vast array of business and government officials and key citizens. Obviously, Sexton intended to build a strong external base before he swung into action. He loved to talk strategy with me, which engendered mutual trust, and then he heaped responsibility on me. Based on Sexton's glowing nomination, I received the staff officer of the year award for the Rocky Mountain Region, a prestigious honor among my peer group. I still consider it one of the nicest honors I've ever received.

Soon after he arrived, Sexton gave me a profoundly useful lesson in the realm of what's possible.

When Sexton arrived, the Sheep Mountain timber sale was in the works but had been appealed by environmental groups. It was located in a gorgeous area just east of Lizard Head Pass along the highway from Dolores to Telluride. Normally, the Forest Service magnanimously discussed all the issues an appellant raised, then reiterated its decision and moved forward, confident that no lawsuit would be filed.

On this occasion, Sexton and I drove to a meeting he had requested with the environmental appellants. En route, he simply said, "I think I'm going to drop this sale."

I was incredulous. The agency just did not do this. Maybe we should not have named it Sheep Mountain, which loomed above Lizard Head Pass as a snow-capped pyramid of Giza. But otherwise the timber harvesting plan was quite reasonable.

This felt to me like capitulating to the enemy. Why not stick with the reliable, "Sorry, but we're right"?

Sexton said he didn't see environmental groups as the enemy. He feared a lot of bad publicity if we pushed ahead, and possibly good will and greater trust if we backed away. He believed we could find other, less controversial timber to sell elsewhere.

Sexton also wanted employees to understand that environmental concerns sometimes had merit. Sexton thought we could change—that we should change.

I realized I had become so captured at times like this by agency dogma that I couldn't think outside the box.

Changes did come to the San Juan. Reality continued to nibble away at the amount of timber offered for sale. Jim Powers, working with foresters who tried to find suitable logging volume over the next few years, made the case to Sexton that previous harvesting practice in much of the timber base made it unrealistic to achieve the planned harvest level of twenty-four million board feet. This was happening all across the country as national forests found it difficult to achieve timber sale goals based on FORPLAN predictions. Sexton further reduced timber harvest to sixteen million board feet.

And timber wasn't the only resource issue. Near Pagosa Springs, the Forest Service caught a rancher leasing his grazing rights. This was contrary to his permit, which required that he graze only livestock he owned. Sexton canceled the grazing permit, which was worth several hundred thousand dollars.

Seldom did the Forest Service exercise this "nuclear option," instead usually invoking a less drastic action like suspending grazing for a year or two. In Bill Sexton I witnessed a forest supervisor with an appetite for bold action.

These actions signaled that the Forest Service had begun regarding environmental concerns much more seriously. There had long been a perceptible tension growing between resource protection and resource extraction. In my reckoning, the resource protectors existed largely outside the agency, while the Forest Service was perceived to be resource extractors. Most leaders I interacted with in the Forest Service would strongly disagree with this characterization, wanting to be seen as astride a white horse, but many rank-and-file professionals thought it accurate.

No such ambiguity existed in our sister agency, the National Park Service. National Park personnel maintained a strong culture as protectors of spectacular national assets, which resonated strongly with public perceptions of their role. In contrast, the Forest Service's conservation mission envisioned wise use of resources. The agency culture that emerged after the Second World War did not strongly identify with safeguarding basic resources as our primary responsibility. Rather, we regarded national forests as working landscapes. The agency's fundamental duty was to produce.

But what about that nettlesome word wise preceding use? That tells me that taking care of the resources comes first. My head and my heart were turning. I witnessed and participated in actions that had the feel of a big schooner heeling over to tack to a new course.

A Glimpse of the Future

John Kirkpatrick had roots in the Southwest Region (Arizona and New Mexico), as did his old friend John Bedell, the supervisor of the Carson National Forest. About 1986, these two instituted an annual triforest meeting of leadership teams, to include the Rio Grande National Forest and its supervisor, Tom Quinn.

These meetings included field trips and discussions on issues of the day—everything from heap-leach gold mining to bear hunting to grazing issues. The teams made good professional connections and enjoyed stimulating talk. The Carson bunch hosted our second event, and we headed indoors for a presentation in Farmington, New Mexico after a couple of days outdoors.

The presenter was Chris Maser, a former researcher with the Bureau of Land Management (BLM) from Oregon, who, near as I could tell, now devoted himself to writing and lecturing about his forest research. Maser gave a lengthy and very detailed lecture and slide show about how forests of the Pacific Northwest function.

I sat enraptured as Maser talked about old-growth forests, squirrels and their poop, fungal spores, mycorrhizae (fungi that grow alongside tree roots to dramatically improve a tree's uptake of water and soil nutrients), lichens, soil-dwelling insects, forest productivity, and, by inference, how just about every aspect of our forest management approach was naïve and simplistic—in a word, wrong. Forests were incredibly complex biotic systems, Maser said. Simple notions of "just managing trees" actually involved significant risk of upsetting the delicate balance of important forest relationships.

I found myself aware that Maser was challenging my own simple notions. This made me both irked and curious.

To begin with, I was ignorant of virtually everything he was talking about (also true, I suspect, for other Forest Service leaders in the room). Maser made a compelling case. If it was true, his conclusions damned the Forest Service. Supporting management with credible science should be the stock in trade of any competent natural resource agency. Yet how can you apply what you do not know or understand?

And so, here I sat in a Forest Service meeting, my head exploding and my heart aching with shame. Many wanted to disagree with Maser, but the credibility of his scientific evidence disarmed them.

I realized that I had relied on what the agency spoon-fed me in various training sessions, and, sadly, that I'd spent little time exercising my responsibility in the professional discipline of independently pursuing knowledge. John Bedell, who was moderating the conference, enjoyed the role of provocateur and relished Maser working our brains with an eggbeater. A wild-eyed discussion ensued about the implications of Maser's message.

I went back in my memory, reviewing the bumps and struggles I'd experienced during the past few years just trying to do my job. I thought about things I'd seen that didn't add up. The worn-out grazing lands on the Bighorn. The ferocious pressure to "get the cut out" even when the forest couldn't sustain it. The gadfly environmentalists that seemed to delight in strewing our path with lawsuits. Now, I recalled, there was a controversy brewing in the Pacific Northwest over a shy, innocent spotted bird.

I realized these issues weren't random and unconnected, but part of a larger reality that we were beginning to perceive. They were like icebergs—not especially menacing until one fully considered the larger threat below the surface. Internally, the Forest Service heard recurrent, insistent voices from many of our best and most credible resource specialists that the agency was ignoring important science that pointed to environmental problems. I'd seen for myself examples where our management wasn't matching our science.

Could that be the reason for the public's growing disenchantment with how we were managing their treasured national forests? Were we that far out of sync with what people, especially some of our own agency experts, believed and valued?

I found myself drifting toward increasing doubt about the merits of our logging policy and practice on national forests. What did the Forest Service owe the timber industry? The general public? The agency seemed to bow and scrape to a largely ungrateful commercial interest, even as the timber industry's power and influence diminished sharply. Furthermore, strong similarities existed in the agency's relationship with a ranching industry whose livestock grazed vast areas of public lands in the West.

What was of ultimate value to the public whose land the Forest Service was privileged to manage? I didn't think the general public viewed commerce on national forests as our top priority. But if they wanted changes, what were we to become? I wasn't certain how to answer that, but I sure thought it had something to do with the Forest Service managing lands with a stronger, more humble environmental ethic.

A Consideration of Values

I considered the implications of the values question through some other miscellaneous, and most interesting, duties, which included public affairs. I hired Ralph Swain, an energetic wilderness devotee with a degree in marketing, and Ann Bond, who had a background in TV news reporting and in hospital marketing, when opportunities arose. These two people influenced me greatly.

I persuaded John Kirkpatrick to allow Swain and me to launch a nonprofit association, Friends of the San Juan National Forest. These interpretive associations have been around for a long time, generally associated with selling nature books and other tourist items at federal and state park visitor centers. Swain and I shared a vision of a San Juan Association (SJA) that would also serve to market the forest and its varied recreational opportunities—if only we could come up with a business model that was financially feasible.

We hit upon a golden idea: forest maps! Every Forest Service office sold national forest maps for a dollar, a more than fair price, I thought. Swain and I talked with Laura Stransky, our receptionist, and asked how many San Juan forest maps we sold each year. She said, "We sold thirty-eight thousand last year."

I asked, "Why one dollar?"

Stransky said, "I don't know. It's just what we charge." I discovered after a little snooping around that we ordered maps from the Denver office's cartographic section at about seventy-five cents apiece.

Swain and I hatched a plan to have SJA sell the maps for three dollars and keep the profits. This gave SJA about seventy-five thousand dollars in start-up revenue the first year, with enough left over to hire a director. We had to beat back an attack from staff in the regional forester's office who claimed we couldn't do this. When Bill Sexton arrived, he said, "Carry on. I've got your back." He waded into the fray to protect this germ of an idea.

Durango had become a national hotbed for biking enthusiasts, so we also developed a biking trail guide as our first marketing project. The guide profiled about thirty-five popular bike routes, and local business sponsors could buy a page of advertising. The association sold these guides to generate additional revenues, while providing a helpful, valuable product.

Meanwhile, Sexton agitated for more TV and video. He said, "Video is the future!" (remember, this was the 1980s) and the Forest Service "is doing nothing."

Sexton and I talked with Ann Bond about igniting action in this arena, taking advantage of her background in TV. With Bond's contacts and production help and SJA funding, we created a twenty-five-minute video called *San Juan Adventure*. It featured the San Juan's incredible opportunities for world-class biking, whitewater boating, four-wheeling, and blue-ribbon trout fishing. The video might look crude by today's standards, but it had enough professional luster that we could actually sell it with a straight face. Our biggest market turned out to be real estate agents who sent it to prospective buyers to sell them on the grandeur of the Four Corners.

Creating a local nonprofit association had symbolic as well as practical value. The Forest Service worked cooperatively with the association to produce revenue, which could, in turn, finance other products desirable to people interested in national forests. SJA could also do things the Forest Service could not. Ideas led to other ideas. SJA was galloping.

We could capture a little cash with these innovations. But even when adding in all our campground revenue, the amount of money we derived from recreation pursuits, such as campground fees, was still far less than logging and grazing receipts. When I considered how many people vacationed in southwest Colorado and how much money they spent on travel, food, and lodging, it seemed clear the total value of sightseeing, hiking, fishing, and hunting was enormous. They didn't pay the Forest Service cash, but their experience didn't come without cost. Clean water and air? Again, no cash changed hands, but these resources had exceptional value.

I didn't consider this to be a brilliant insight, but it seemed the Forest Service managed lands in a way that noncommercial resources were severely undervalued and underrepresented. And much of our public was getting fed up.

Swain, Bond, and I conceived a new target. The beautiful 23-mile highway from Silverton to Ouray has long been known as "The

Million Dollar Highway." In today's economy that wouldn't build a mile of road, but the name did have a catchy ring. There was an even longer stretch of highway—a 236-mile loop that included the Million Dollar Highway—connecting the towns of Durango, Silverton, Ouray, Telluride, and Cortez. This stunning stretch begged for some creative marketing. Attractions like the Durango-Silverton Narrow Gauge Railroad, Telluride and Durango Mountain ski resorts, and Mesa Verde National Park hung like pearls along this road, which traversed some of the most spectacular scenery in the nation.

Bond and I thought the magnetic power of this road, if marketed properly, might draw more people to southwest Colorado. The communities could work together rather than compete with each other. And the Forest Service held the key, since most of the route coursed through national forest landscapes. Of course, we sought to attract more people to their national forests and improve the quality of their recreational experience.

This route deserved a name. SJA sponsored a name-the-highway contest that attracted about sixty entries from all over the country. We asked entrants to create a name and a logo. A panel of judges selected "San Juan Skyway" and its simple, evocative logo as the winner.

From this simple beginning, great things ensued. Building off spectacular scenery and abundant four-wheel-driving opportunities, we now had in hand a 236-mile-long necklace adorned with recreational pearls. We dove in with local communities to market and publicize the merits of the Skyway, while improving the driving experience with thoughtful interpretive improvements of its abundant historic sites and scenic vistas.

The Federal Highway Administration had a pool of funds derived from a portion of gas taxes to enhance auto touring. Dick Ostergaard, a Forest Service landscape architect, eagerly utilized these funds to design and develop several scenic turnouts over a period of a few years. Communities along the route began to cooperate, pooling tourism funds (from Durango's pillow tax, for example) to market the Skyway imaginatively, allowing all towns along the way to benefit.

Newspaper and magazine articles quickly followed, highlighting a variety of great recreation opportunities, such as an autumn Aspen Tour for motorcyclists and the Alpine Loop for four-wheeling, hiking, mountain biking, and ski touring in winter. Travel books featured the San Juan Skyway with splashy articles. And about this time, the Forest Service launched its Scenic Byways program featuring beautiful drives all over America, and designated the San Juan Skyway as the first. In my opinion the Skyway remains the best.

I've come to believe that these few examples illustrate that we sought to respond to a profound truth, perhaps without full understanding. The great worth of our national forests is grounded in environmental values like scenery, clean water, abundant fish and wildlife, and unfettered recreation opportunities. Although the Forest Service historically devoted most of its budget and staff to commercial activities like logging and grazing (hence its being an agency in the US Department of Agriculture), the public most craved opportunities to experience personally the land they owned in a largely natural condition.

I began to understand the distinction between dollars and values. The Forest Service might not be generating revenue from the vast suite of environmental services these lands provided, as we did from logging and grazing, but environmental services flowing from forested, mountainous landscapes will always far exceed the value of logging and grazing. In fact, they are priceless. And nature provides these gifts to us humans free of charge. Stewarding public lands must respond to this essential truth or it will fail.

Washington, DC, Calling

At a planning meeting near Estes Park, Colorado, Larry Larson from the chief's office approached me about a couple of vacancies there. I didn't see DC as a good place to work because of the cost of living and the risk of being trapped in DC without a guaranteed return ticket to a national forest assignment.

Larson assured me that his staff group had a great track record of moving beyond DC to plum jobs, if they did good work for him. Did I want a challenge? But these were peripheral issues. I most feared losing connection with the land if I went to work in the "puzzle palace," as some people referred to the chief's office.

I conferred with several confidants, including John Kirkpatrick, now in the chief's office. Kirkpatrick confirmed that Larson's planning group enjoyed a sterling reputation, and assuaged some of my other fears. I was forty-three and thought I might need more experience to advance further. Without strong conviction, I filed all the necessary paperwork. Things stayed busy at the office. I didn't think about it much more.

My office looked north out of a third-floor window of the Federal Building in downtown Durango, up the Animas Valley toward Silverton and the Needles Mountains. About mid-July every year, monsoon season arrived. Thunderstorms rolled across the San Juan almost daily. Lightning peppered the ridges. Great gray sheets of rain drooped from the thunderheads. The usually brief storms had incredible intensity.

One particular day, I sat enjoying the show when Bill Sexton asked to see me. He had a job offer for me from Larry Larson. Boom! Sobered, elated, scared all at once. I said I'd need to discuss it with my wife, then I'd let him know. An unwritten rule governed applying for jobs in the Forest Service—don't apply for a job unless you intend to take it.

I walked the short distance back to my office, closed the door, and let the tears flow. There would be no more Animas Valley lightning shows. My experience in Durango had been truly eventful. Three children birthed here, one of them—Veronica—stillborn and buried. A resurrected career. Deep friendships at work and church. A deep connection with Four Corners country born of joyful fishing and biking. I still consider the San Juan to be among the most beautiful and spectacular national forests in the United States. Leaving would prove hard, but soon we were bound for DC.

I held in my soul the stirring of a land ethic, new to me. I did not know what it would grow into.

Chapter 5: Land of Wonder

WASHINGTON, DC, PART I, 1989–1991

Confronted by the enormous escalator at Rosslyn Station, I quickly decided to attack. My stay in temporary quarters in Alexandria, Virginia, necessitated a bus ride to the Pentagon, then a descent into DC's metro system for a subway ride north to Rosslyn Station, then the ascent to daylight and a short walk to the Forest Service personnel shop to fill out some paperwork. Fresh from Durango's sixty-five-hundred-foot elevation, I figured the escalator presented a good challenge. Soft bureaucrats stood to the right while I took the tough-guy route to the left. I didn't just walk up. Sprinting in a suit would have been unseemly, but I put some hop in my step.

Halfway to the top, I was still going strong, cruising as the escalator succumbed to my determination. Three-fourths of the way up now, my heart pounded and my legs felt the burn. I was still in "I think I can" territory. At the top, unwilling to yield to common sense and my arrogance in full control, I crossed the finish line to no applause.

As I strode toward the office in the spring humidity, a powerful flush struck me. Sweat blossomed from every pore, quickly drenching my just-pressed shirt. I beat a hasty mental retreat, promising to never do that again.

And I didn't. I recall a great bumper sticker I saw in Wyoming: "Eat sheep! 10,000 Coyotes Can't Be Wrong." Going with the flow has certain legitimacy. I learned the hard way not to race up a long escalator. Learning hard lessons became a pattern in DC.

I arrived in DC in 1989, a pivotal time. A decade of excessive timbering on national forests closed with a discernible shift in agency actions. The environmental movement blossomed and began to noticeably affect and influence Forest Service thinking and policy. The movement spawned and then helped shape the attitudes of a new generation—including many Forest Service employees who were often specialists bold enough to question agency dogma.

Environmentalists had gone from being perceived as crazy to credible. The Forest Service frequently lost lawsuits, especially related to protection of endangered species and water quality. "Enviros" had the agency on its heels. Few Forest Service leaders dared consider themselves environmentalists, but they now paid careful attention to those who were.

The Big House

The Forest Service, like most of the USDA, was housed in the South Ag Building—a monstrous cream brick structure big on space, short on looks. I settled into a roomy office with Ann Loose, my new associate in our planning group. I quickly learned that the building's blueprints came from the federal penitentiary at Leavenworth, Kansas. Kind of scary, our office being the perfect size for four inmates. As I looked around, I could picture one bunk bed on each side and bars on the window. Good ideas, even good people, could become imprisoned here.

Another forest planner, Vern Fleischer, from Utah, and I had just joined the staff as appeal coordinators. National forest plans were being completed all over the country, and many were being appealed by any person or group with a bone to pick and a postage stamp. We worked in a fashion similar to a law clerk for a judge. We analyzed each appeal, reviewed the adequacy of the Forest Service's written record, and developed our findings in a written decision that eventually became the agency's final resolution. If the appellants wished to contest the matter further, they would have to take us to court. We worked very closely with the USDA's Office of General Counsel to ensure that our decisions had a solid legal footing.

We refrained from any *ex parte* contact with our field organization. The appeal process required that we confine our analysis to the written record, such as environmental analysis and official decision documents, so as not to be influenced by any internal persuasion. However, we did coordinate with various staff groups in the chief's office. For example, if the appeal dealt with an endangered wildlife species, we worked through the issue with wildlife specialists. If the appeal involved logging issues, we worked with the forest management staff. All agency parties involved in the appeal rigorously reviewed each decision draft to present to the Forest Service judge—usually Dave Unger, the associate deputy chief for national forests, but occasionally George Leonard, the associate chief—for final editing and signature.

Larry Larson, my immediate supervisor, mothered us like a hen, directing us to just the right people, helping keep us on task and on time. Larson ensured that decisions made good sense and adhered to policy. Poor work reflected badly on him, and he knew it could also hurt my reputation.

My first appeal review involved the Superior National Forest plan in Minnesota's far north. Environmental groups claimed the Superior's timber harvest was excessive and that the timber program wasted money. Larson asked that Denny Schweitzer, our staff economist, review my draft. Ouch! My draft came back with abundant red ink. Schweitzer supported my conclusions, but the quality of writing "lacked distinction." I sensed another escalator experience coming. Schweitzer's review was not personal. He said I just had to do better work. Plenty of people could help me learn the rigors of crafting a well-reasoned, articulate argument. I needed to lose any pride of authorship and work hard. "Can you do that?" I didn't think no was a good answer.

My two and one-half years as an appeal coordinator proved a great experience in critical thinking and critical writing. I believe I became sufficient—good enough at reasoning and writing to make a coherent case for a sensible decision.

I also was given assignments other than writing appeal decisions. These proved interesting and educational because they gave me a glimpse of a variety of internal conflicts. One afternoon I was told in hushed tones to watch the broadcast of ABC's *20/20* that night because Chief Dale Robertson's interview with Chris Wallace "had not gone well." Several agency staff, including me, had been selected to watch the broadcast and come to work at 6:30 am the next morning to discuss spin control.

Never had I been so eager to watch *20/20*. The show did not disappoint. Chris Wallace, feisty, pressed the chief hard on issues. I rooted for the chief to represent the agency well, but I suspected that most viewers thought Wallace was getting the better of Robertson. Was it true, Wallace queried, that the Forest Service had squandered millions of dollars intended to reforest timber harvest areas?

The camera squared fully on Robertson's face as he stared at Wallace. An embarrassing silence followed and then just mushroomed to overwhelm the scene. It seemed an eternity until Robertson grinned awkwardly and gave a weak response. The interview proceeded downhill from there. I thought, "So this is why I'm watching." Surely this was the "not gone well" part. I squirmed at the chief's obvious discomfort and my own. This was far worse than I'd been led to believe.

The Forest Service employs well over thirty thousand people, and a great number of them would be watching this program. Our agency culture held the chief as one of us, and we all wanted him to represent us well and proudly. The chief's performance in the interview struck me as painfully disappointing, and, frankly, embarrassing to the agency as a whole.

I eagerly headed to work the next morning, composing, in transit, my own humble, thoughtfully worded mea culpa from the chief to the entire organization—an honest admission that he'd let us down.

About eight of us showed up at the assigned time and place and waited for Susan Hess, the public-affairs director, to lead the discussion. Curiously, she did not show. After quite a long time, we crafted a plan as we waited for Hess. Someone had taped the interview, so we watched it again. And again. We timed the "Great Silence"—six full, agonizing seconds of dead air. If anything, our first impressions had been too charitable.

Because most of our employees lived in the Mountain and Pacific time zones, we knew the chief could limit the damage by crafting a quick electronic message to all employees. It would contain an admission that he'd not done well representing the agency, quickly convey what he should have said, convey his trust and confidence in us, and conclude with a strong reminder of our great mission. Get it done and mailed out no later than nine am. We needed to make this message the first thing that all employees would see when they came to work. Create a good buzz, manage the coffee talk.

Susan Hess finally showed up a bit after nine am, two and a half hours late, and beyond the nine-am spin-control opportunity. Hess confronted a very antsy and agitated group, but she was happy, all smiles. She had just come from the chief's daily briefing, where the executives had all agreed that nothing need be done. We could just go back to work after she uttered a quick thanks for our being willing to help.

Pandemonium! We were incredulous. What had they been smoking? How could any Forest Service employee watch the debacle on *20/20* and conclude everything was cool?

We pressed Hess hard that the chief owed the agency something. She left with parting assurances that she'd discuss our concerns with the chief. Late that day, about four pm, the chief distributed an electronic message to all employees about the *20/20* interview. Just mush. I thought it actually made things worse.

I learned later, by rumor, about the truly horrible reality of the interview, unseen on TV. Hess had allegedly broken into the interview "on camera" insisting that it stop immediately in hopes of stanching the bleeding. Providentially, the electric power crashed. People cooled off in the hall while the technicians restored lights and the interview ensued. The two-hour taping was boiled down to a fifteen-minute segment. A consensus emerged that the really bad stuff had been left out.

Thus did our agency pooh-bahs express relief that no great damage had been done. Keep calm and carry on! Stiff upper lip and all that.

Our group did our best to convince Hess that the chief had no clothes, and we thought it best to admit this. We did not prevail. Our group had been asked to help. We spoke truth to power. What seemed so clear to us seemed to escape the bosses completely. I sensed they were utterly out of touch with our field organization. Hell, they were out of touch with reality!

I continued to plug away on forest plan appeals. A unique case crossed my desk that reassured me we could deal with knotty problems and take a strong stand for the environment. Forest Service processes can and do yield good results.

Hoosier National Forest in southern Indiana issued a plan that allowed no off-highway vehicle (OHV) use at all. The Hoosier's plan made it the only national forest with a no-OHV policy.

This did not sit well with Tom Lennon, the OHV specialist in the recreation staff group, who seemed quite concerned. The Forest Service caught hell from the four-wheel-drive lobby, and we ultimately got sued by several prominent off-road vehicle organizations.

My review of the lengthy appeal record revealed a clear, well-reasoned decision. The forest supervisor and regional forester had both concluded that the Hoosier National Forest best served the public as a bulwark against noisy vehicles running rampant on public land. One key criterion in their decision was that the Hoosier, like most eastern national forests, had numerous private land inholdings nestled in and around federal forestland. Farmers and rural residents had spoken loud and clear that they enjoyed their quiet bucolic country living and expected the Forest Service to manage public lands so as to protect their interests and property. OHV enthusiasts, for their part, made the claim that they had an implied right, as citizens, to recreate as they wished on public lands.

The plan carefully assessed all the pros and cons and concluded that a no-OHV policy best represented the broadest public interest. Period. The decision leaned heavily on a 1972 executive order issued by President Nixon that clearly spelled out the necessity to consider and address negative impacts of OHVs, such as noise and trespass, on neighboring lands.

Although their conclusion was unusual, the Hoosier officials had made a convincing argument in denying OHVs. Lennon, the OHV expert, and I, the appeal coordinator, would have our day in court before Associate Deputy Chief Dave Unger, the official who would decide the appeal. Lennon viewed the prohibition as generally unfair to OHV enthusiasts and also thought it would set a bad precedent. I concluded that the Hoosier plan represented a reasonable decision, correct given the local circumstances, and defensible among several available options. The forest supervisor had exercised his discretion appropriately. Unger agreed and supported the Hoosier's decision. We also prevailed in court, and Hoosier National Forest remains the only national forest that excludes all OHV activity.

Truth Speaks to Power

In February 1989, Jeff DeBonis, a forester on Oregon's Willamette National Forest, wrote a bombshell letter to Chief Robertson. DeBonis's broadside directly confronted the consequences of timber harvesting in the Pacific Northwest—clear-cutting, old-growth forest liquidation, degraded water quality, depleted salmon fisheries, and endangered species. DeBonis stated that agency practices and policies could not be justified or sustained, and he claimed the Forest Service supported the "narrowly focused, short-sighted [timber-industry] agenda to the point that we are perceived by much of the public as being dupes of, and mere spokespeople for, the resource extraction industries." DeBonis was "talking about corporate greed versus a priceless national treasure."

The letter created an immediate sensation. It was combative, yet respectful; resolute, but not whiny. It constituted an unmistakable and direct affront to the chief and to prevailing Forest Service dogma.

Remarkably, almost nine months passed before DeBonis received a reply, Chief Robertson noting that "I just have not had time to focus on the many ideas you included in your letter."

Really? I found that remark absurd and so disappointing. How could the chief have "not had time to focus" on the most compelling issues facing the agency? Robertson told DeBonis his letter was "one-sided and did not appropriately take into account the interest and needs of the American people," and that it was important to have a "more balanced view of the world under our multiple-use mandate" (no doubt referring to the Multiple-Use Sustained-Yield Act of 1960[9] relating to timber, range, water, wildlife, and recreation management on national forests). The chief's response was a patronizing pat on the head.

DeBonis created a new nonprofit organization—the Association of Forest Service Employees for Environmental Ethics (AFSEEE). The title alone noted that the agency lacked environmental ethics. I'm guessing DeBonis knew this to be a career-limiting action. Although he remained with the Forest Service briefly, he ultimately resigned to devote himself fully to building AFSEEE, which remains viable today as Forest Service Employees for Environmental Ethics (FSEEE), scrutinizing the Forest Service and dutifully pointing out when the agency's environmental ethics are lacking.

DeBonis's letter touched a nerve with many employees who had a growing unease with agency policies that seemed decidedly pro-tim-ber. Many new professional staff such as wildlife biologists, fisheries biologists, botanists, landscape architects, archeologists, and hydrolo-gists—some called them "combat-ologists" —had been employed by the agency to help assess the consequences of timber harvesting. Many of them became increasingly concerned about the negative consequences of logging and skeptical about whether the Forest Service valued their opinions. They grew weary of making good-faith recommendations that were either ignored or tweaked to ease the movement of logs to the mill as usual.

DeBonis, a forester and former Peace Corps volunteer in El Salvador, agreed with many of the disaffected specialists. DeBonis bundled their frustration and gave it voice.

Other important red flags had been rippling in the wind. In the mid-1980s, Chief Max Peterson convened a leadership summit for for-est supervisors held at Utah's Snowbird resort. I still remember John Kirkpatrick returning to Durango, enthused about Snowbird's theme: "Traditional Values, but not Traditional Methods." Seeking to build on the success of Snowbird, another leadership meeting, dubbed "Sun-bird," convened in Arizona in 1990. Virtually every forest supervisor attended, as well as the entire top tier of agency leaders.

Conflicts simmered and then boiled over. Almost half of the forest supervisors had signed a tough letter to the chief, arguing the agency was off the rails. DeBonis represented the thinking of a midlevel forester, but this letter came from the very heart of agency leadership. Timber programs continued to command an enormous share of the budget, even though public constituencies clamored for a higher priority on recreation, wildlife, and water quality. The forest supervisors concluded that the Forest Service was not living up to its credo, "Caring for the Land and Serving People."

Another group of supervisors brought a video that showed several of them pleading for a reduction in timber harvest and a reordering of priorities. Both the letter and video quickly made their way throughout the agency and into the public square. This was an insurrection, no getting around it. First DeBonis, and now this.

The renewed publicity and scrutiny stunned and angered Chief Robertson and his cohorts. Thankfully, no retribution followed. Perhaps so many offenders made it impossible to single out particular individuals. But DeBonis and the forest supervisors initiated a critical, introspective look into the soul of the Forest Service. Agency policies might become sharper on the whetstone of public values.

A Fuzzy Future Appears

Deputy Chief Jim Overbay asked Hal Salwasser, the Forest Service's national director of wildlife and fish (later a regional forester and research station director as well as dean of Oregon State University's College of Forestry) to lead a small task team that would look into Debonis and Sunbird with this double focus: "What's going on? What should the Forest Service do in response?"

Something important blew in the wind. In spite of Robertson's dismissive reply to DeBonis, here was evidence that the Forest Service sensed it was increasingly out of step with the times. Although the questions for Salwasser's inquiry were vague and portrayed insecurity, they touched on the values shift I felt while on the San Juan. The sentiment echoed throughout the country as many leaders groped for some kind of meaningful response. Change was afoot.

Thus, when asked to serve on Salwasser's team, I jumped at the chance. Hal Salwasser impressed me as being as intellectually sharp as anybody I'd met in the Forest Service. His ability to critically evaluate information and develop solutions stood out dramatically. I assumed the opportunity to work with Salwasser on any project would bear fruit for both the Forest Service and me.

We were to function under the oversight of Overbay, for whom I'd first worked years ago when he was supervisor of the Black Hills National Forest. He had risen steadily to this important post. Overbay gave our team carte blanche and asked that we dig deep. He wanted to hear whatever we had to say, and he promised to stand by our work when we were done.

The team—I don't even think we had a title—met and talked. We talked a lot. Unlike many teams that developed tasks and objectives, we functioned in the philosophical realm, basically thinking out loud amongst ourselves to discern the root causes of unrest and the implications for how we accomplished the agency's mission. We tried to discern what was really going on. We strove for simplicity. We developed a quite straightforward message: the Forest Service behaved as if forests were about trees but had neglected the profoundly bigger picture that management policies should also effectively consider wildlife, fish, water, soil, scenery, and, frankly, everything that made a forest a forest.

Attitudes and values and beliefs and ideology shape action. We concluded that the agency acted in ways that were increasingly at odds with the public's environmental ethos and the voice of the land itself. The Forest Service needed a New Perspective. This title strongly suggested the need for a new attitude, rather than a slate of specific actions. I liken this to a heart transplant—if we could get our heart right, action would follow suit.

Our team prided itself on articulating a simple, evocative idea, but not pride in the sense of having invented something. Rather, we felt as if we'd lassoed a wild horse and gotten it tied to a post where it could be closely inspected. Extending the metaphor, Overbay mentioned after a briefing that he didn't quite know what to call the horse, but he thought the Forest Service should ride it.

We were emboldened. The time came to present our findings to the chief and his leadership team in a packed conference room. Decorum demanded that we let Overbay and Salwasser do the talking. Overbay provided the context, and Salwasser delivered the content. Of course, they did a very capable job. It seemed to me the chief and other leaders grasped and understood our message. Then Chief Robertson spoke. He agreed with our team's general concept of the New Perspective, but he had misgivings about the scope of the concept.

The chief noted that "we are the Forest Service," directly responsible for management of national forests. Robertson thought it best that the application of this so-called New Perspective be narrowly confined to forests therein.

As that pronouncement lingered in the air, my mind raced, considering all that I thought hung in the balance. First, the Forest Service considered itself to be a leader in conservation, yes, even globally. If we'd identified something fundamentally transformational (I thought so), it could have, should have, global implications.

Second, we managed grasslands as well as forests, fish and wildlife habitat as well as trees. What about the agency's research branch, the world's largest forest research organization? New Perspectives ought to inform and guide research as well.

And what about our State and Private Forestry organization, which provided financing and technical support to state foresters and all manner of private forest owners? Leaders should boldly articulate how a New Perspective ought to influence management of all aspects of forest environments to address the entire suite of environmental values and services beyond merely trees.

Then, after a brief moment while Chief Robertson allowed people to ponder his view, he said—and I remember this clearly—"Does anybody disagree with that?" This question displayed both humility and a challenge. The chief's willingness to entertain a difference of opinion was inviting, but he had taken a position, and he was the chief. Better not open your mouth now unless you had your act together.

I waited for Overbay or Salwasser, especially, to adroitly point out the broader implications of New Perspectives and wisely suggest to the chief that he and the Forest Service ought to square ourselves and embrace a bold, expansive vision.

I contemplated speaking from the far corner where I stood among other onlookers. The golden moment shone briefly, hanging tantalizingly while the chief scanned the room.

When you're at bat in baseball, you have a split second to decide whether to swing. Many thoughts follow the trace of the ball after it leaves the pitcher's hand, crowding the time it takes to arrive at home plate. Decide! Swing or take the pitch. Then the moment passes.

Suddenly, Robertson said, "All right, then. We'll move forward on that basis. Thanks for the hard work."

Over, and so quickly. I stood up, ashamed of my reticence to speak. I wanted someone else to be the one who spoke up. Salwasser, representing the team and our product, should have said something, but maybe he, like me, feared taking on the chief. I've rolled that moment over in my mind for many years, wondering, "What if?" Maybe nothing would have happened. I do believe it would have at least sparked a good discussion. And maybe, just maybe, New Perspectives, writ large, might have become a much broader vision.

But no. Now we had another problem. As we began to take New Perspectives for a walk in the real world, it proved difficult to grasp and was too ethereal and touchy-feely. It was difficult to describe succinctly, hard to measure, and too philosophical. New Perspectives didn't enjoy the warm embrace of the chief and even alarmed some people.

We decided on a new name that strived to be more practical and descriptive. Thus did the term *ecosystem management* come about.

I don't recall now who suggested it, but ecosystem management had more substance and heft. It implied doing something, more action than thought. And it clearly suggested something much more comprehensive than timber management.

Adopting ecosystem management as a guiding principle did not change everything instantly. Yet it did capture a fundamental shift to a broader, deeper, more comprehensive understanding of natural resources that I believe will continue to govern management of forests for the foreseeable future.

The Nine Lives of Clear-Cutting

Clear-cut logging had persisted as a controversial practice for decades. The only thing that angered environmentalists even more than clear-cutting involved clear-cutting of old-growth timber. An appeal assignment came my way for the Cherokee National Forest plan, situated in Tennessee, home state of then-Senator Al Gore. Gore, on a Senate committee that dealt with Forest Service business, was no fan of Chief Robertson. (Robertson noted in his memoir, while watching the nominee Clinton name Gore his running mate at the Democratic Convention, that he turned to his wife and said, "I'm a goner if Clinton's elected." And so he was.)

An environmental group had appealed the use of clear-cut logging on the Cherokee. I received a note to meet with the chief to discuss the appeal. The request was highly unusual because appeal coordinators rarely met one-on-one with the chief.

I went to Robertson's office, a spacious suite on the fourth floor of our new quarters, the splendid, red-brick Auditor's Building across the street from the South Ag Building. Robertson enjoyed elegant views of both the Washington Monument and the White House. The chief said he wanted to make a statement with the Cherokee appeal. He planned to write the clear-cut section of the appeal decision himself.

And, by God, he did. He called me to his office first thing the following Monday morning and laid a hand-written three-pager on the table. He wanted me to read it quickly. He busied himself while I took a few minutes to scan through his write-up. I worked through his meticulous account of the agency's history with clear-cutting as he reasoned his way toward a conclusion. How riveting to read the rationale of the chief forester of the United States! He surprised me by declaring that "clear-cutting is not an appropriate practice in scenic mountainous areas." Thus, he had determined that there would be no further clear-cutting on the Cherokee.

The chief had chosen to personally intervene in order to use this appeal to get the word to the field organization that he, the top leader, had spoken definitively on the issue. He discussed the need to send the decision to all line officers, calling specific attention to the section proscribing clear-cutting. Time to take a stand! Hopefully, Senator Gore would take note also.

I said I got all that, but (remembering my regret about remaining silent on New Perspectives) felt I needed to mention that many national forests, perhaps most, would be considered "scenic mountainous areas." Like, for example, the Cascades in the Pacific Northwest, which was the agency's timber breadbasket. Clear-cutting persisted there as standard practice. Might the chief want to tighten up the wording a bit to provide more specificity, remove the sweeping generality? No sense in painting ourselves into a corner.

Quickly, with a grin, he said, "The Cascades are different!"

I knew Chief Robertson had intimate knowledge of the Cascades, having been supervisor of both Siuslaw and Mt. Hood forests in Oregon. I also couldn't believe my ears. I didn't want to argue, but I thought a good staff person should point out something that was likely to hit the fan. Should I press the issue harder or just move on? Was I confusing fear and respect? I thought I had made my point clearly and believed the chief understood me but just disagreed. I bit my lip and left the office. The decision went out immediately, just the way the chief had written it.

There's more to the story. President George H. W. Bush's administration had pressed the Forest Service hard for a "substantive action" to announce at a United Nations environmental conference in Rio de Janeiro in 1991. Bush wished to reveal to global partners his serious intentions to protect the environment. The Forest Service quickly came up with a sweeping seven-point policy letter that, for all intents and purposes, doomed clear-cutting on national forests. The swift, decisive action taken by the agency impressed me, and, I guessed, almost everybody in the Forest Service. Cherokee! Rio! It seemed the chief was stepping up to the challenge, signaling the likely end of clear-cutting on national forests.

Less than a year later I conferred with Bob Devlin, director of timber management for the Pacific Northwest Region, about the agency's prohibitions against clear-cutting, noting that it appeared the Cherokee decision and Rio policy letter were having no effect whatsoever on clear-cutting activity—still "same old same old." I didn't sense people knew, or even cared, that these strictures existed.

Devlin kind of laughed dismissively, as though curing me of my naïveté, and said, "Those are just policies. They're not really binding."

So that's how things work? Flummoxed, bewildered, and deflated, I couldn't even find words to respond.

To this day, I still wonder—"Who's the chump here? Me or the chief?" Did he make the policy actually expecting clear-cutting to stop? Or did he assume he'd be ignored? After quite a bit of reflection, I've concluded that the chief really intended his policies to have full force and effect, but that he was alarmingly out of touch with a field organization that winked at his edicts. How could an agency be so cavalier that gaining compliance necessitated treating field officers like children?

Either way, policy in action (or inaction?) seemed fundamentally dysfunctional. As for me, did I just not get how to play the game? Is it a game? A hall of mirrors and deception? Such things confound my sense of truth.

New Perspectives. Ecosystem Management. Clear-cutting. Engaged on important issues, I had opportunities to shape agency policy, though it remained frustrating that my work yielded so little in results. But I felt fortunate to be involved.

Dave Unger, the associate deputy chief for whom I often prepared appeal decisions, led an ad hoc old-growth task team and asked me to join the effort. The old-growth team's accomplishments remain vague. We did have good, honest discussions that recognized and attributed intrinsic value to old-growth forests, finally catching up with the findings of our own researchers and debunking the oft-heard notion of old-growth forests being biological deserts. The work of scientists like Jerry Franklin, Fred Swanson, and Chris Maser showed quite the opposite. We developed assessments of the amount of old growth on federal lands and trends in the loss of this important asset. We struggled to agree on practical measures to protect remaining old-growth forests. We sensed the Forest Service would take no dramatic protective actions, even knowing of the dramatic decline in old-growth forest acreage.

At one point I walked with Unger toward his office, making a philosophical plea for bold action and wondering aloud as to what held the Forest Service back. Unger patiently, respectfully offered that "incremental decision making" is an agency's usual practice, rather than sweeping reform. He asked me for patience, thanked me for my enthusiasm and commitment. That helped take the edge off my attitude, but it fueled a desire to be in a position to do more.

I wondered if I'd look back in twenty years and see any progress in protecting old-growth forests. At this rate, it seemed unlikely.

A Higher Calling

I'd assumed I would pursue staff work after my stint in DC, most likely in a job as planning director working for a regional forester who managed all forests in several states. Yet a number of office mates transferred out to prominent line officer jobs as forest supervisors. As I listened to each of them describe the hunt, I sensed a hunger growing within me for a job that I considered beyond my ability.

I had traced the history of one controversial issue after another on forest after forest for the past two years. Inevitably, I wondered how I might address each situation if I had been the forest supervisor. What if a different choice had been made, might a different decision have negated the need for an appeal altogether? The arguments, especially those of environmentalists, began to make more sense, more often. Most forest plans I'd evaluated seemed pro forma, lacking imagination and evidence of a strong land ethic. The vision articulated for national forests in plan after plan seemed too small for the greatness of the landscapes. I considered the respect and admiration I had for leaders like Kirkpatrick and Sexton. What might it be like to manage, to lead an entire national forest?

A growing conviction welled within me. I strongly believed the Forest Service needed to manage lands with greater concern for the environment. I yearned to put into practice my growing convictions that we could do better at aligning our practices with public values.

This was my turning point. Leading a national forest to pursue exemplary land stewardship now seemed desirable, attainable, and I hungered for it.

To be considered a viable forest supervisor candidate involved a delicate dance in the nuanced culture of the Forest Service. The only path to selection lay in gaining the trust of Deputy Chief Jim Overbay, who personally approved all selections. Making the list of potential forest supervisors seemed akin to the Romans' thumbs up/thumbs down vote. I needed to discern if I had Overbay's support, but the situation called for a tangential approach.

Bob Jacobs had replaced Larry Larson, who was off to Utah, as my new boss. Jacobs had been a forest supervisor in Florida. Overbay had likely handpicked Jacobs to lead the planning shop. Jacobs and I got on well. He respected my work, and we enjoyed discussing resource policy. Jacobs told me he thought I had the potential to be an effective forest supervisor and that he would support me.

Approaching Overbay about Furnish-as-forest-supervisor, Jacobs received a warm response. But not hot. Jacobs suggested the time had come for me to discuss that possibility with Overbay directly.

I made an appointment with Overbay. I can still feel those butterflies as I walked into his office. Very friendly, Overbay explained that I was one of many well-regarded candidates in the chief's office. He noted that I had done well in my job, turned a few heads, and would probably get my shot. DC experience still counted as almost essential to be a good forest supervisor, and this alone accounted for many aspirants doing time in DC. But the top of the list right now held higher-priority candidates than me. The chief and Overbay himself would ensure that capable DC candidates were selected for a reasonable percentage of upcoming vacancies. My disadvantage was that regional foresters made recommendations, and I was hardly known outside Colorado and Wyoming. If I sought a job elsewhere, Overbay would likely have to convince a skeptical regional forester that I was the best choice.

I watched over the next few months as individuals were picked. Jacobs sat in on these meetings, and he was very kind to keep me posted. "Your name came up; you need to start applying for open jobs. Don't be picky. You need to signal that you're willing to serve anywhere. Understand what I'm saying? You have my support. You're getting close. You were in the top three on that one."

I applied for several jobs throughout the country—Minnesota, Utah, Washington, Oregon, Wisconsin, New Hampshire. I found the updates as each job was filled to be exciting and intoxicating, almost more than I could bear.

Then I journeyed to Milwaukee for a face-to-face interview, which was the norm there but unusual in other regions—most selections were based on the paper application and one's reputation. But I'd never worked there and was unknown. Jacobs told me that Butch Marita, the regional forester, had a home and property within the Chequamegon-Nicolet National Forest in Wisconsin, the very national forest I was vying for. Marita had the reputation of a meddler, and Jacobs said I could anticipate his being in my hip pocket if I got the job. That alone would make the job more of a challenge.

I also applied for a few deputy forest supervisor jobs, including on the Gifford Pinchot in Washington and the Siuslaw in Oregon. It was October, and my wife, Judy, and I sneaked away for a quick vacation in Maine to enjoy the brilliant foliage. I left contact information with Jacobs in the event he needed to reach me. Judy and I dined at Graziano's, a favorite Italian restaurant in Lisbon Falls, Maine, and then lounged back at the motel. The phone rang. It was Bob Jacobs. I locked eyes with Judy, the moment electric. This was it. Jacobs wasted no time. "I have an offer for you!"

I was thinking Wisconsin—close to my folks in Iowa, they're getting older, it will be nice to see them more often. The Great Lakes States are nice. How different this will be from all those years out west. Could be my last job, maybe retire there?

"Deputy job on the Siuslaw."

I know I spoke as quickly as possible and tried to sound enthusiastic, but a moment's hesitation hinted of disappointment. Judy and I both knew it could now be years more to become forest supervisor. So did Jacobs. Maybe it would never happen.

"Jim, I can tell you're a little disappointed. Don't be! This is a fantastic opportunity and a great forest."

I knew this was true, and it was time to say yes and thank him for all his help in making this possible for me. A delicious blend of joy, relief, wonder, and excitement washed over me. It was special, too, to share this moment with Judy, who had walked this long path with me.

The regional forester in Portland, John Lowe, was formally notified of my acceptance via personnel channels. Wendy Herrett, my new boss, then called to welcome me. She had been the first woman district ranger in Forest Service history. She blessed me with the gift of being chosen. I will always be grateful for her trust.

We had one car left after donating the hand-me-down Caddy from Judy's folks to the National Kidney Foundation. I never saw myself as the Cadillac type. The December cross-country trip was an opportunity to visit my folks, seldom seen now, in Iowa.

After a couple of rich, celebratory days with them, we made the long drive to Fort Collins, Colorado, in one day. I dropped Judy and the kids off for an extended visit with her family. I drove on alone, the gray Honda humming the tune of the road spanning the bleak winter landscapes of Wyoming, Utah, and Idaho. Finally I crossed over the Snake River into eastern Oregon. I'd had ample solitary time to ponder the opportunity before me. Cresting the snowy Cascades west of Sisters, I entered the home stretch in late afternoon.

I'd heard much about the winter rain in western Oregon. Coursing alongside the Santiam River, I passed through Sweet Home, Corvallis-bound, then encountered scattered showers. A dark gray sky loomed over the lush, vivid green of the Willamette Valley's grass-seed fields. Oregon was wet, yes, but mild for the dead of winter. This land seemed calm and soothing, a real contrast with the Puzzle Palace I'd left behind. Maybe, I thought, I'll soon enjoy some of that peace and quiet. How wrong that was.

Chapter 6: Intractable Conflict

Siuslaw National Forest, Part I, 1991–1999

Wendy Herrett graciously welcomed me to Corvallis, the Siuslaw National Forest, and my quite splendid deputy forest supervisor office, right next door to hers. I'd never met Herrett, but I knew her by reputation. The Forest Service chose her, a landscape architect, as the first woman district ranger, first in Meeker, Colorado, then in Hill City, South Dakota. She'd served as deputy forest supervisor on the Mount Hood National Forest immediately before taking the forest supervisor job on the Siuslaw. Although she was not the first woman supervisor, she was among the first few. She considered all forest resources to be vitally important, a distinction that threatened the Siuslaw's timber-first legacy and, as I came to learn, sowed seeds of discontent.

The Siuslaw National Forest, at about a thousand square miles, is relatively small. I know many ranger districts that are larger than this entire forest. Yet my research when applying for the job had told me that the Siuslaw was the seventh largest national forest in the United States in terms of staff and budget—disproportionately large because big timber drove the agenda. Oregon's Coast Range consists of coastal temperate rainforest and grows trees like nowhere else, owing to abundant rainfall, deep soils, mild temperature, and low elevation.

The weight of being on such a significant forest in big timber country began to sink in. One of the timber staff group dropped in asking that I sign a letter seeking about fifteen thousand dollars in trespass damages from a timber company.

"What's this about?" I asked. He told me that we'd surveyed a property boundary and discovered three trees that had been illegally cut by a logger. He said, "This happens all the time." He'd already called and told them the letter was coming. I'm thinking, "Five thousand dollars per tree?! This is a different world than my familiar Rockies."

As the first few weeks passed, I discovered the reasons I ended up in Corvallis. I'd succeeded Dave Garber, a wildlife biologist who moved on to become forest supervisor on the Gallatin National Forest in Bozeman, Montana. Herrett acknowledged that she and Garber did not make a good team. He had wanted to leave, and she had sought a deputy who wasn't wedded to a timber-driven ideology, and who better matched her style and leadership agenda. She carefully evaluated the applicants and came to believe that we'd work well together. She made a point to tell me, "We need to be a united front." This would prove so true.

Snails, as they make their way to wherever they're going, proceed slowly with antennae extended. If they touch anything, the antennae recoil quickly. I can feel my antennae fully extend when I'm beginning a new job. I am very sensitive to clues like office decor, staff clothing, facial expressions, and friendliness of coworkers as I try to feel the vibe. Every office has a culture, a unique personality. The Siuslaw culture quickly struck me as sober, very competent, and also a bit beleaguered. We had outstanding professionals in most every staff group. But cheerful? No. Interest groups contested almost everything the Siuslaw did.

I wanted to get into the woods right away to see examples of Oregon's Coast Range forestry. This would help me get grounded as to how the Siuslaw did things. I asked the timber staff officer, Dale Rettman, to give me a look-see at what the Siuslaw had going on with the New Forestry I'd heard so much about in presentations back in DC. Some field foresters, I'd heard, were trying to apply some of the research Chris Maser had described by doing "dirty" clear-cuts—so called because, instead of the clean, antiseptic look of a forest stand where all trees (even dead ones) were cut down, these clear-cuts had numerous live and dead trees left for wildlife, stream protection, and visual diversity. Big dead logs on the ground stayed put, too.

Rettman kind of smiled and said, "We're not really into that, so I don't have anything to show you."

I asked him if we at least had some timber sales in the planning stages that intended to apply New Forestry principles.

"Nope. This is the Siuslaw. We clear-cut here."

A bit puzzled, I tried to discern the attitude at work in this place. I didn't sense any embarrassment or guilt or even mild shame. No, it seemed more to be pride, perhaps even a touch of arrogance—no need to waste time on a passing fad.

Being the new guy from DC, I asked Herrett if she would object to my hosting a get-together in Corvallis, and she readily green-lighted my request. I wanted to share my views on Ecosystem Management and so-called New Forestry. I needed to hear staff on the Siuslaw share their thoughts; even more so given Rettman's comments. I extended an open invitation for a one-day dialogue in Corvallis to any and all who wished to attend.

I was pleased, maybe a bit surprised, to see about seventy-five people filling our conference room. We had a lively discussion for several hours. People shared their interpretations of Ecosystem Management and how they felt about it and what they thought it would look like on the ground. Pointedly, they wanted to know what I thought, what my views were. I shared that I had helped develop the New Perspectives program, as well as serving on the agency's old-growth team, and that I had a keen interest in the debates that had erupted pursuant to the DeBonis controversy and the challenge to agency leadership that had emerged from the Sunbird meeting in 1990. I said I detected a very strong aroma of change in the air. I believed the Siuslaw to be geographically and politically at the very core of these issues.

I assumed that most people knew of the friction between Herrett and Garber, and they probably wondered whether I would become a supporter of Herrett and where she wanted to take the Siuslaw. With so many people in the room, I took the opportunity to mention that I was anxious to help her achieve her goals.

Tony Vander Heide, our planning staff officer, had supervised me briefly when we both worked on the Black Hills planning team. I recalled Vander Heide as progressive, competent, intense. The Forest Service had selected Vander Heide, a hydrologist, for a subsidized master's degree in resource planning from Michigan State in about 1975. This was standard practice within the agency to help promising employees further their education with the aim of increasing agency expertise.

He had come to Corvallis after graduate school and had been here since, recently having shepherded the Siuslaw through a ten-year forest planning process. Most people credited him, quite fairly, with being the driving force and brains behind the Siuslaw forest plan. I wondered how it would pan out for us working together again.

Vander Heide invited me for a day trip to get a look at the Siuslaw, with my promise to hear his take on the forest plan and his thoughts on present circumstances and the future, including his relationship with Herrett, of course. He wanted me to understand the Siuslaw's hard-earned and well-deserved reputation as perhaps the "best timber forest in the country."

The weather dealt us a typical winter day, low gray clouds with intermittent rain. Oregon's Coast Range contains some of the most productive forest in the world, measured in terms of timber yield per acre—an AAA rating, so to speak. This inherent productivity, coupled with years of aggressive forest management practices, had produced tremendous historical timber yields.

Vander Heide took pride in the forest plan. The timber harvest level of 215 million board feet per year had been a major accomplishment. Oddly enough, the accomplishment rested on reducing timber harvest expectations, because the regional forester in Portland had wanted more timber—something a lot closer to the historic level of 320 million board feet. The reduced harvest level derived from the firm insistence by Vander Heide and Tom Thompson, the forest supervisor at the time, on a seventy-year rotation age rather than the commonly used forty-year industry standard in the Coast Range. Cutting trees when older would, of course, reduce timber yields. Vander Heide told me there had been a pitched battle with the regional forester in Portland, who controlled all national forests in Oregon and Washington and wanted the forty-year rotation age and the higher timber volume. The Siuslaw, and Vander Heide personally, had prevailed. But the Siuslaw was still a timber forest, no doubt about that. Vander Heide thought it essential that we commit to the forest plan as the best course for the future.

He frankly stated his belief that Herrett lacked commitment to the timber program. He thought Herrett seemed wobbly and unwilling to persevere with all the heavy lifting necessary to accomplish even the reduced harvest he'd fought so hard for. He questioned whether she had the leadership capacity—the guts, frankly—to meet the forest plan's commitments. He refused to stand idly by if she failed to adhere to the forest plan.

I said those sentiments seemed to put him squarely at odds with Herrett. He acknowledged as much, but said somebody had to fight for it. I sensed he saw Herrett as a foe.

As Vander Heide and I sat in the car overlooking the domain of the Siuslaw, ridge upon ridge unto the far horizon, I felt I could see the future he described. My eye focused on the numerous clear-cut units, intrusive as they were upon the vast sea of green forest. Now it was 1992, and we were thirty or forty years into converting this landscape into a tree farm. What I saw before me contrasted sharply with the natural forest born of fire that we had inherited.

The region's history indicated that severe fires had burned virtually the entire mid–Coast Range in the late 1800s. I imagined hot, dry winds out of the east that licked at careless homesteader fires in the Willamette Valley to erupt and blow all the way to the Pacific Ocean. When established in 1907, the Siuslaw National Forest held little allure for investors looking for commercial timberland. They passed by the steep, forbidding country with its young, dense forests. Better lands lay elsewhere.

So it was left to grow, and, in time, a great forest emerged. When demand for timber soared after the Second World War, the Siuslaw's big trees became very profitable. Technology conquered the steep slopes and deep draws with clear-cuts and new roads, so that heavy logging blanketed the land. I caught glimpses of remarkable coastal old growth in the rare pockets that had been protected from fires, and even the relatively young (yet deemed "mature") 130-year-old timber was magnificent, with huge trees and numerous streams that might host a hoped-for recovery of salmon. As a publicly owned forest, the Siuslaw had the potential to stand in sharp contrast to nearby industrial forests dedicated to maximizing money.

I said, "Tony, it makes me sad to think what this forest will become if we keep doing what we're doing. This forest has much greater potential than that."

A chasm opened between us. He clearly understood that I stood with Herrett and not with him.

I went on to remind him that forest supervisors have a lot of power. Did he have a plan to overcome her inherent advantage as his boss? He said, "I plan to wait her out."

"Interesting plan," I said, "but what if she stays here for a long time?" He said, "I don't think she'll be here long." His hunch proved correct. Maybe it was no hunch.

Herrett was aware that many of her leadership team openly opposed her, and she identified Vander Heide as the most prominent antagonist. His behavior bordered on insubordination at times. Given Vander Heide's concerns, this did not surprise me. At our monthly meetings I witnessed evidence of Herrett's concern firsthand. Acute tension hung in the air, even open hostility at times. After meetings, Herrett's eyes often teared up when she described how hurt and frustrated she felt. I felt awkward, not knowing quite how or whether to comfort her, yet appreciative of her trust in allowing me to see her vulnerability. I resented her being bullied and seeing how this hurt her.

She refused to show tears in front of others, fearing this would be read as weakness. How true. Women in leadership positions remained relatively scarce. Herrett mentioned to me more than once how she felt she must "act like a man" to be seen as an effective leader. I thought she had a stiff backbone, but often went about accomplishing her objectives in ways, yes, foreign to me. I still had to grow in understanding that women are effective, though embracing a different management style than many men. Cultural transformation came slowly, came hard, as many new leaders emerged who were not self-acclaimed "tough foresters." She feared that displays of emotion would only confirm the stereotypical belief of her as a weak woman, and thus incapable of strong leadership.

How had Herrett and her opponents on our team arrived at this juncture? The Forest Service has a strong culture based on a deep affinity for natural resources. There are many employees who share a view that forest resources are best used in service to our country. Many other employees believe the Forest Service has abused its conservation mandate with excessive resource extraction, necessitating greater environmental protection and restoration of degraded landscapes. The larger public generally cleaves along this same fissure.

Vander Heide and Herrett, archetypes for these groups, respectively, now vied for power and supremacy. I saw both of them as dedicated professionals, each advocating different and passionate environmental ideologies. How does one determine the best ideology? Surely not on the basis of brute power, whether exercised from above or below.

Herrett did not want to continue the past practice on the Siuslaw because of visible and factual evidence that it was not ecologically sustainable. When natural forests were logged, the wood they produced often came at the expense of endangered species. We had come to realize that logging destroyed habitat for spotted owls, and we would soon understand that salmon and marbled murrelets also were at risk. Furthermore, persistent controversy and lawsuits—which the Forest Service was losing—attested to strong public opposition.

It seemed clear that we needed a course correction. I agreed with Herrett that present circumstances called for pursuing a different destination, which meant, at minimum, reduced timber harvesting and enhanced wildlife and fish protection. She may not have known exactly how to get there. But who did?

A few months after my arrival, she and I conducted midyear performance reviews for each district ranger and staff officer. It didn't take long for me to tell her what I thought was really going on. Forget about numerical ratings and adjectives, I said. Our leadership team's dynamic was a mess, and our effectiveness was near zero.

I named three or four individuals and said, "These guys are just killing you, Wendy." I told her that it would be difficult for her to advance her vision in the face of such deeply entrenched opposition.

I saw it as basically Us versus Them. I thought we should use the midyear reviews to assert a powerful united front. We needed to convey our position that their official year-end performance rating would suffer unless we saw changes.

She appreciated my support but said she just didn't have the stomach for escalating the battle. She thought this would make things even nastier. I acknowledged that it would very likely get messy, but that dire circumstances dictated strong action. She preferred to just gloss over the contention and live with it as best we could. I feared an already bad situation would continue to deteriorate.

At summer's end, we heard that Chief Robertson selected Lyle Laverty, the recreation director in Portland, to be his national recreation director in DC, a big promotion for Laverty. Within hours of this news, Herrett abruptly asked me into her office. She said that John Lowe, the regional forester, had called to ask her to replace Laverty. I wanted to congratulate her but for the tears. She said she "felt like such a failure." John Lowe did not ask her to consider this, in spite of a bit of sugar coating. She was being told, in Forest Service–speak, that she needed to step aside and take the job in Portland. In her mind, the message strongly hinted that she presented less of a problem in Portland. The inference hung in the air that the Siuslaw would benefit from a change at the helm.

I tried to comfort her, to little avail. I saw before me a valiant warrior, a leader and role model approaching the end of her career, who looked utterly beaten. I felt so powerless. She was gone by the end of September.

Regional Forester Lowe designated me as the acting supervisor, with a 120-day temporary boost in pay. Thank you very much. I considered how helpful the past ten months had been in allowing me time to grow into this challenge. I reflected on being passed over for the top job in Wisconsin. How scary, I thought, that I'd actually felt ready for that job at the time. The time spent learning under Herrett had been a gift. She had graciously tutored me, offering counsel, advice, and explanations of how and why she approached things the way she did. I resolved that I had to be tougher if I were to accomplish anything. Not tougher than her, but more determined, disciplined, resolute, and focused than those employees who I perceived stood in the way of a better destiny.

I arrived at work that first Monday after Herrett left, keenly aware of an unfamiliar and heavy weight on my shoulders. I had spent a couple of hours at the office on Sunday afternoon, the day before, crafting a letter to Tony Vander Heide. I intended the letter to remind him that, although Herrett had departed, I remained. I saw no need to mention that he had indeed waited her out. We both knew his instincts had been correct. But I rebuked him sharply and said I would no longer tolerate disrespectful conduct, toward me or anyone. I put him on notice that I expected his behavior to improve markedly. I signed the letter, carefully folded it, placed it in a blue envelope (meaning this was personal, for his eyes only) with his name on it, and put it on his desk.

I'd come to work Monday ready for the bomb to explode, and it did. Vander Heide immediately came into my office, angrily clutching the letter and demanding an explanation. He wanted to know if this con- stituted a letter of warning, an official personnel action.

I said, "No. It's simply a letter from me to you, informing you of my expectations."

He denounced me, said that no one had ever said such derogatory, disrespectful things to him, and he would see to it that I regretted it. Why didn't I have the guts to give him the letter in person? I actually agreed with that last part.

In the fabric of personal dynamics, there can be a rending such that no mending is possible. That brief, ugly argument in my office marked virtually the last time that we spoke to each other. I did not envision that outcome when I gave him the letter, but the ensuing silence was not entirely a bad thing.

Over the next few months things were ugly and got worse before they got better, at least for me. In the end, three of the four difficult people whom Herrett and I discussed, Vander Heide included, no longer worked for the Forest Service. A district ranger took early retirement. Vander Heide and another staff officer received a directed reassignment (meaning they had to accept the transfer or leave the Forest Service) to distant district ranger jobs that I assumed (correctly) each would decline. Both retired. The fourth and last person took a demotion.

When presented with opportunities to bring careers to an end, I seized each chance. I had been ruthless in ways I thought myself incapable of. I liked myself less because of it, but I acted in the belief that these actions would restore the health of our organization.

Going forward, I relied on the principle that Jack Booth had shared with me many years ago: if the people you have won't work with you, find people who will. As each of the four vacated their positions, opportunities opened for new blood. I recruited capable people, each of whom played a pivotal role in creating a new team dynamic. I think I acted wisely, but this team helped to navigate as best we could while the Siuslaw's timber program plummeted and we hemorrhaged budget and staff expertise. I had no useful corollary on which to build our strategy. We confronted widespread organizational fear that bordered on panic. We faced an uncertain future with a new and untested leadership team. Everything before me lacked clarity.

Yet all my twenty-five-years of experience were available to draw on. I'd had some wonderful mentors. I considered how the various positions and experiences in my career to date honed me to prepare for this challenge and helped me steer through the crisis. Might there be reasons why circumstances had thrust me into this particular situation? I remember Deputy Chief Overbay saying the Forest Service really made an effort to get just the right person in each forest supervisor job. I began to believe that maybe, just maybe, I was the right person for the Siuslaw right now.

I served a lengthy eighteen months as acting forest supervisor due to a government-wide freeze on promotions and transfers. When the freeze lifted, John Lowe told me I'd done good work in the interim and thought I'd earned the forest supervisor job. But there was a catch. When the job advertisement came out in October 1992, right after Herrett's departure, I decided not to apply because I'd been in Corvallis for only ten months, and I thought my candidacy would be seen as presumptuous. Also, seldom was a deputy promoted in place; most moved to a different national forest when selected as supervisor. I welcomed the chance Lowe afforded me to remain on the Siuslaw and pursue our new mission.

But Lowe was nonplused when I told him I had not applied. Though unusual, it seemed we had but one course available: downgrade the forest supervisor position from a GS-15 (referring to the Government Service ranking scheme for federal employees) to a GS-14 and laterally reassign me. I suggested it seemed reasonable because the dramatically reduced timber program resulted in a significantly smaller budget, workforce, and program of work. Lowe agreed and processed the paperwork for my appointment as Siuslaw forest supervisor, effective May 1994.

The events of the next few years restored my faith in people and in nature's capacity for healing and restoration. I remain incredibly proud of the work the Siuslaw's group of leaders did in building a new reality. We pursued a vision that honored Wendy Herrett's convictions.

Chapter 7: From Despair to Hope

SIUSLAW NATIONAL FOREST, PART II, 1991–1999

In 1991, a ruling from federal district judge William Dwyer stopped the Forest Service from cutting timber from forests inhabited by the northern spotted owl. The injunction clamped the jugular of federal timber in the Pacific Northwest and sent a deep shudder through the agencies and the timber industry that depended on these forestlands for raw material. The end had finally come to the decades-long liquidation of the remaining old-growth forests.

It had the feeling of a meat locker door clunking shut behind you...and then hearing a bolt lock into place. Judge Dwyer had spoken. The good old days were gone forever.

Prior to Judge Dwyer's ruling, the Forest Service had inexorably, incrementally sought to tame the vast and valuable forests of the Pacific Northwest. Aggressive logging of public land rested on a foundation of a growing nation's demand for wood. And the nation trusted the Forest Service as long as its policies seemed consistent with public values. But public values can change. Distrust can grow.

Following the war in Vietnam, a new generation viewed government with increasing skepticism and cynicism. Fewer and fewer people accepted sweeping vistas dominated by clear-cuts and new roads.

Instead, they valued naturalness, clean water, abundant fish and wildlife, and a deep sense of connection with the land. They were anguished at what the Forest Service was taking from the forest at the expense of future generations.

You can plant all the trees you want, but you can't make a forest. That's God's business. Environmental groups successfully attacked the traditional Forest Service practice of clear-cutting old-growth forests, using the northern spotted owl as the arrow to successfully pierce the bureaucratic armor. The agency's own planning regulations required it to protect the viability of wildlife, including owls. Judge Dwyer based his ruling on convincing evidence from the plaintiffs that the Forest Service was cutting down the very same old-growth forest habitat that was essential for owl viability. Using the Forest Service's own scientific research that demonstrated population declines, environmental groups proved that federal laws did not allow the agencies to threaten the owl's viability. In this, they scored one of the most significant legal victories in the battle between commerce and the environment.

The Siuslaw National Forest was in the crosshairs of the spotted-owl controversy. In response, it forged a new vision and strategy for managing federal forestlands. This new Siuslaw model uses management approaches that nurture, rather than eliminate, mature and old-growth forests and their associated fish and wildlife habitats. It also reflects the values and aspirations that most people in the region have for their national forests. I say most, because many traditionalists within the Forest Service, and certainly the timber industry, remain opposed.

The Siuslaw became essentially the antithesis of the kind of industrial forestry practiced on national forests for decades. What if all national forests undertook a thorough and fundamental rethinking of how to achieve the basic mission of stewarding public lands in the face of changing values and circumstances, such as climate change?

As supervisor of the Siuslaw National Forest during most of the 1990s, I confronted the wrenching challenge of reframing our mission subsequent to the owl crisis. Yet the necessary changes in forest management produced real triumphs.

Doubling Down

The Siuslaw National Forest, established in 1907, has a history of aggressive timber harvest, mostly in the post–World War II era. Siuslaw timber management mirrored industrial forestry—clear-cut the standing timber, burn logging slash to prepare for planting a dense monoculture of Douglas fir trees, spray herbicides to eliminate competition, and (often) plant genetically superior seedlings, then fertilize to give it all a boost. The Siuslaw spent more money per acre than any other national forest because it generated the most wood and revenue per acre.

About 1990, Siuslaw timber production began to decline precipitously in the aftermath of lawsuits and legislation. Three pivotal factors affected the Siuslaw: spotted-owl population declines; then conflicts over the marbled murrelet, a small oceanic bird that nests in big coastal conifers; and finally plummeting salmon stocks.

I've already spoken of the atmosphere within the Siuslaw management team that contributed to this leadership challenge. An essential first element of problem resolution is problem recognition. The real message from Judge Dwyer was that institutional denial and resistance to changing traditional forest management prevailed at the highest levels of the Forest Service and permeated all levels of leadership.

Judge Dwyer's 1991 opinion[10] stated:

> The problem here has not been any shortcoming in the laws, but simply a refusal of administrative agencies to comply with them. This invokes a public interest of the highest order, the interest in having government officials act in accordance with the law. More is involved here than a simple failure by an agency to comply with its governing statute....The most recent violations of the NFMA (National Forest Management Act) exemplifies [sic] a deliberate and systematic refusal by the Forest Service and the Fish and Wildlife Service to comply with the laws protecting wildlife. This is not the doing of the foresters, rangers, and others at the working levels of these agencies. It reflects decisions made by higher authorities in the Executive branch of government.

Former Forest Service chief Dale Robertson was among those found by Dwyer to have deliberately and systematically refused to comply with laws protecting wildlife. Robertson had been forest supervisor of both Siuslaw and Mt. Hood national forests early in his career. Remarkably, Robertson assumed leadership of the Siuslaw at the very young age of thirty-one, indicating that superiors recognized his great potential. Robertson, highly respected throughout the Forest Service, continued to rocket up through the agency and became chief at forty-seven. He and other leaders knew the Pacific Northwest forests well. I think they consistently maintained timber harvest at high levels in the well-intended belief that this policy best served the interests of the public and the agency.

I could see that the policies and actions of these leaders laid bare a deep schism between the past and the future—forest management as it had been done and what it needed to become. The goal for decades had been to maintain the annual timber harvest total for all national forests at a level of about eleven billion board feet. Thus, Forest Service leaders managed owl issues to minimize disruption of timber production. Timber harvest became even more important than legally required protection of the owl and other wildlife. The lawsuit successfully demonstrated that this policy was both illegitimate and illegal. A different approach was needed.

The same schism evidenced itself throughout the nation at the field level. Siuslaw managers issued their forest plan in 1990, just before I arrived, having taken almost ten years to complete the effort. The scientific basis for protecting the owl had started with researcher Eric Forsman's 1976 master's thesis on the northern spotted owl, completed while he was a graduate student at Oregon State University. Forsman had become interested in the owl after observing them in the Corvallis watershed, which lies along the eastern fringe of Siuslaw National Forest and is a source of drinking water for Corvallis. He described the critical link between late-successional forests, also called old growth, and the owl's habitat needs. Eliminate old-growth forest, his research suggested, and you eliminate owls. In 1987, the Forest Service hired Forsman as a research wildlife biologist at the Corvallis Forest Sciences Lab.

If Forsman's research had an embedded message for the Forest Service and its timber management practices, one would think it would have been heard most clearly in Corvallis. Yet dedicated adherence to timber production marked the fifteen-year period between 1976 and Dwyer's judgment in 1991, both on Siuslaw National Forest and throughout the Pacific Northwest region.

Siuslaw managers released their forest plan just prior to Dwyer's injunction; the owl controversy had become white-hot. The plan established an allowable timber sale quantity of about 215 million board feet (mmbf) per year, a reduction from the 320 million-board-foot annual sale level throughout the 1980s. Many thousands of jobs and billions of timber dollars hung in the balance. Although the owl issue had already begun to threaten traditional timber policy and harvest levels, the Siuslaw plan made a very clear statement that timber occupied the seat of highest priority, and that clear-cutting of mature timber would continue. Minimal measures—the least possible— would be taken to protect the owl.

Numerous other plans for national forests in owl habitat followed the same thinking as the Siuslaw's. The Forest Service was doubling down on its bet against owls.

But Dwyer's injunction dramatically changed all this. As the new Siuslaw forest supervisor in the immediate aftermath of the injunction, I observed widespread anger and avowed defiance. Yet a small number of employees expressed philosophical agreement with Dwyer, pleased that the obsession with timber harvest might possibly be coming to an end.

Indeed, when Wendy Herrett became forest supervisor in 1990, she tried to refocus the Siuslaw's mission away from timber and toward other forest values, but she encountered intransigence from many employees who regarded her with deep suspicion, even hostility. Herrett seemed almost relieved that Judge Dwyer had imposed a decision on the Forest Service. Herrett sought to carefully navigate through this unstable environment, recognizing, inevitably, that a very different future lay in store.

The spotted-owl controversy emerged as a key environmental issue in the 1992 presidential election. I cannot recall another Forest Service issue that figured so prominently in a presidential race. Candidate Clinton visited Oregon, then a swing state, saying that he would fix the owl problem. President-elect Clinton, making good on this campaign promise, convened a one-day forest conference in Portland, Oregon, on April 2, 1993, with heavy regional news coverage. Almost all the president's cabinet attended, listening all day long to sincere and desperate pleading from all sides that had a stake in the outcome.

The President noted in his opening remarks[11], "The worst thing we can do is nothing. As we begin this process, the most important thing we can do is to admit, all of us to each other, that there are no simple or easy answers. Let's confront problems, not people."

In the months following Clinton's Portland appearance, the leadership group of Siuslaw National Forest had engaged in a discussion of our resource management purposes for this landscape. Our conversation culminated in a vision statement we called Siuslaw Stepping Stones. I have never been involved in a more rancorous and passionate clash over resource ideology. Our capacity for unified leadership was clearly fractured. The Stepping Stones vision did not enjoy strong consensus and received an ambivalent welcome. Yet, I think this exercise helped the Siuslaw's leadership team prepare for the sweeping changes that resulted from the forest conference because we had already contemplated a dramatically different future.

The planning effort to develop the owl solution, the Northwest Forest Plan (NWFP), commenced immediately. A draft plan was published in July 1993, an astoundingly swift, ninety-day effort. The final NWFP, a sweeping new vision for federal forest management, was issued in April 1994. I thought it quite similar in concept to our recent Stepping Stones vision.

Briefly, the NWFP accomplished two aims: it satisfied Dwyer's demand that vertebrate species populations remain viable (a legal requirement of NFMA), and it also struck a new balance point between owl habitat protection and timber production by sharply reducing logging on federal lands inhabited by owls. The reductions took the harvest level down from about four billion board feet per year to less than one billion board feet.

The NWFP was huge, complex, and radically different from the management plans it replaced. When I first read it, my reaction was "You gotta be kidding me!" Many of its foundational concepts were theoretical and untested. As a transformational forest management model, it placed an enormous burden on our field organization to make profound and swift changes to how we did business.

The Journey Begins

I vividly recall announcing the NWFP decision to a large group of Forest Service and Bureau of Land Management employees at Corvallis's Majestic Theater. I began my remarks with a simple rhetorical question: "Do you believe this plan can work?"

I thought this was the central issue because it focused on personal will. The NWFP had been crafted by research scientists, unlike all the individual forest plans that had been developed by local forest managers. Most managers resented the idea that researchers had usurped their prerogatives to dictate how their forest should be managed. Managers also assumed the new plan would be an unwieldy, impractical, unworkable guide. These factors fomented skepticism. There seemed to be remarkably little commitment to make the NWFP work, or even to give it a chance.

However brilliant the NWFP might be, field personnel had to believe in it or it would not work. I suspected many, perhaps most, employees thought the NWFP flawed, deficient, and unworkable. I assumed the work of implementing the NWFP would be difficult, yes, but it merited my best effort. So I answered my own question: "Yes. I believe the plan can work."

The NWFP reduced Siuslaw timber harvest from 215 to 23 million board feet, a 92 percent reduction that far exceeded even my most severe prediction. This dramatic change held dire implications for our workforce, exacerbating the steep decline due to repeated budget reductions. Timber had been the lifeblood for the Siuslaw. This level of reduction would mean hundreds of our staff would lose their jobs.

I hated this part of the job. I'd never faced this kind of workforce reduction, but I knew it meant heart-wrenching choices for me and bitter hardship for those affected. As optimistic as I tried to be, this was going to be a gloomy time.

Moreover, few agency leaders outside the planning team itself had an opportunity to methodically consider the natural resource implications of the NWFP because of the secrecy surrounding the plan before its release. In general terms, the NWFP imposed three zones on the forest landscape. The first zone was late-successional reserves (LSR). These consisted of forests that would be left to grow into old-growth habitat. The second zone consisted of riparian reserves (RR). These were streamside areas about four hundred to five hundred feet wide, intended to protect important terrestrial and aquatic habitats. No commercial logging was to be allowed in either the LSRs or the RRs.

The third zone was called the matrix. It consisted of whatever productive forestlands were left after the first two zones had been designated. Timber would be harvested only from matrix lands, and then only if harvesting didn't endanger other wildlife and plant species.

Because of the character of its forestlands, the Siuslaw National Forest was dominated by LSR and RR, with only about 6 percent of forestlands remaining in matrix. Our leadership group tried to ferret out the implications. We now knew that nine out of ten mature timber stands had nesting owls and murrelets—which meant no more timber harvest. Clearly, it seemed a losing proposition to continue to invest effort and large sums of money in traditional clear-cut harvests of mature timber. Cynics thought the Siuslaw targeted owl habitat for destruction, but it was more the case that the birds used virtually all the available mature forest habitat. If we cut big timber, we necessarily bumped into birds.

Furthermore, harvesting in the matrix would be subject to what the Forest Service called "survey and manage" protocols, which were intended to detect the presence of other sensitive, old-growth-associated species, such as certain mollusks, lichens, fungi, and red tree voles (prey of the spotted owl). And on top of that, we had to think about the dramatic reduction in wild fisheries over the past decades. It was urgent that we try to restore fish habitat on the Siuslaw's many streams.

The Forest Service has navigated its way through numerous controversies in the past century, but few have been as complex, layered, and consequential as the northern spotted-owl issue. The NWFP rationale for drastic reductions in timber harvest was based on the fact that large-scale clear-cutting of old-growth forest profoundly affects fish and wildlife habitats. The convergence of owl, murrelet, and salmon habitats in the Oregon Coast's temperate rainforest coalesced to bring Siuslaw timber production to a virtual standstill.

Although the NWFP actually allowed for clear-cut harvesting in matrix areas, the plan did not require harvest. Given the high likelihood that owls and murrelets would make it illegal to harvest anyway, our leadership team concluded that common sense and economic efficiency dictated a new direction.

Then we discovered another nasty little wrinkle in the analysis. The NWFP used a relatively narrow standard for riparian reserve width. This had caused us to overestimate the amount of matrix land we had. We recalculated the area of matrix land using the proper width for riparian reserves. We found to our dismay that the additional land base removed from matrix resulted in a planned annual sale quantity of only seven to nine million board feet. We alerted the regional forester in Portland that the planned sale quantity of twenty-three million board feet per year specified for Siuslaw National Forest needed to be revised downward even more.

In sum, the plan's standards and our experience with its habitat protection provisions made it clear that any further pursuit of clear-cutting in mature stands in the Siuslaw's matrix would prove fruitless. I regarded the NWFP as an explicit admission that this incredibly productive landscape could not simultaneously maximize both wood products and wildlife. The forest was the womb that sustained this natural abundance. Thus, the NWFP made the hard choice for federal land. The remaining mature forest in the Coast Range would stay standing.

Building a New Boat

Economics was really the only motive for cutting mature timber. Most stands in the Coast Range originated in the mid- to late-1800s and could be characterized as being in the prime of life, since they often lived to be four hundred to five hundred years old. Biologically speaking, no need or urgency existed for logging mature timber.

Our tree plantations, however, were another matter. We knew of about 120,000 acres of young stands that had grown up on sites that had been clear-cut starting back in the 1950s. The practice of that day had been to clear-cut, remove all the old snags and dead logs on the ground, burn the site, plant dense stands of Douglas fir seedlings, and then spray the sites with herbicide a few years later to reduce competition from deciduous trees. Many of these young stands were, or soon would be, of sufficient size to support commercial thinning.

These plantations were not natural in the same sense of coastal forests that naturally regenerated themselves following the large, infrequent, severe fires that have swept through the Coast Range for the past hundreds or thousands of years. I thought about it this way: when you cut and remove all the standing trees, then plant only one species (Douglas fir) to densities that exceed four hundred trees per acre, these trees ultimately exclude almost all other plant life. And that was the point—to grow Douglas fir because of its economic value and eliminate the competition. The plantations had a destiny: clear-cut again at age seventy and start the process all over again.

So there was a legitimate concern that these tree plantations would not attain the desired mature-forest character evident in natural stands without further management. Having been created and managed solely for timber production, the plantations had far too many trees per acre, and they also lacked diversity.

The intent of the NWFP was to manage both late-successional reserves and riparian reserves to ultimately attain old-growth characteristics. On the Siuslaw, however, the young Douglas fir plantations within our late-successional and riparian reserve areas might never become old growth without intervention. I viewed this as a critical issue. We decided to focus our available budget and resources on management of these young stands, but not for wood production. We decided to thin these plantations, rather than clear-cut at age seventy, to reorient them toward a new destiny as old-growth forest.

Our simplified strategy would allow the mature forests in the Coast Range to age and become old growth in time, without our intervention. We focused our management instead on the many young stands peppered throughout the LSR and RR to improve the likelihood that they too would some day provide high-quality old growth.

We knew our thinning strategy would also have significant financial consequences. The stumpage value (value of standing trees on the stump) of logging mature forests often exceeded forty thousand dollars per acre in 1994 when the NWFP was issued. Thus, the value of standing timber on this relatively small forest easily exceeded ten billion dollars. In contrast, thinning a young stand might yield only a few thousand dollars per acre.

Many environmentalists familiar with the management and history of the Siuslaw remained skeptical and distrustful. We worked closely with both Oregon State University and the agency's Pacific Northwest Research Station to establish a credible scientific premise for our strategy.

My team and I quickly identified harvest priorities in young plantation stands, largely in our late-successional reserve areas. A research project, Coastal Oregon Productivity Enhancement (COPE), helped our progress significantly.

COPE, a ten-year research partnership supported by money from government, industry, and interest groups, was housed at Oregon State University. As a COPE board member, I sensed we had a unique opportunity to do thinning research. Stu Johnston, a forester at Mapleton Ranger District, designed a study to evaluate three different levels of thinning on plantation sites that had more than two hundred trees per acre. The thinning study called for taking out most of the young trees, but leaving stands with one hundred, sixty, and thirty trees per acre, as well as establishing a control site where no thinning would occur.

We used a traditional commercial timber-sale process to achieve the treatments, awarding the contract to a high bidder. The goal of this study and several similar ones then being conducted was to see which thinning treatments work best in encouraging a young stand of trees to grow into the old-growth forest characteristics that certain wildlife need for their habitat.

The scientific rigor of COPE research, along with several field seminars conducted by the partnership, established that alternative management approaches like these really could work in young forest stands. As a result, we gained credibility with a variety of interest groups—federal regulatory agencies such as the US Fish and Wildlife Service and the National Marine Fisheries Service, state and federal forest managers, industry foresters, and environmental organizations. These stakeholders now recognized that thinning young stands in late-successional reserves and riparian reserves could achieve mature-forest conditions.

In 1994, just after the release of the NWFP, I delivered a speech to the public and agency employees. I titled it "Uncle, Thank You, Please." The speech sent a message to the public and agency employees that the NWFP openly admitted a need for change ("Uncle"), that we owed a debt to environmentalists for their passionate insistence on principles of sustainable stewardship ("Thank You"), and that we needed time and patience to perform up to new expectations ("Please").

The speech enhanced the willingness of environmentalists to allow Siuslaw leadership some latitude to demonstrate our sincerity and commitment to change. *High Country News* (*HCN*), a biweekly news-paper "for people who care about the West" requested my permission to reprint the speech, and I gladly consented. *HCN*, well respected and widely read, offered a huge media footprint for people who cared about national forest issues. I received congratulatory notes from around the country, although, unsurprisingly, the speech outraged some of those who still clung to the traditional timber-first approach.

We quickly adapted Siuslaw timber-sale programs to rely almost entirely on young-stand thinning. In addition, we pioneered many simplified administrative procedures to reduce costs and increase revenues. The Forest Service commonly uses some of these today in administering its stewardship contracts.

Looking back over those tumultuous times, I can identify one pivotal event that crystallized the moment our forest organization made the leap from the old to the new. Our leadership team was in a rather heated discussion about where to go with our timber program. Some were trying to persuade us to retain as much of our traditional activity in mature forests as possible, while others thought we should acknowledge that was a dead end, that it was time to pursue thinning. We needed a decision.

I had heard enough, and my guts told me to take a stand. I said, "We will no longer clear-cut mature timber on the Siuslaw." We would devote all our energy to thinning plantations. This was our future for as far as I could see; we had decades of this work before us.

Wrapping up the Roads

As we were reinventing the Siuslaw's timber program, our road system transitioned from an asset to a liability. Past road construction, while costly in the Coast Range, had proved manageable because the timber sales were earning more than enough to cover it. By 1990, Forest Service road mileage exceeded 2,700 miles, or 2.7 miles of road per square mile, much of it paved or high-standard graveled roadway. These roads and dozens of bridges had cost hundreds of millions to build and now required maintenance costing about $4.5 million per year. We'd been using both regular budget funds and cooperative deposits from timber purchasers to do the maintenance. As timber harvest and appropriated funds both fell, our road maintenance capacity dropped by about 75 percent.

What's more, forest roads, especially older ones, are prone to landslides, which are hard on streams, fish, and wildlife. I ordered a complete reassessment of our road management. I could see no way forward that would not require us to close a significant portion of our road system.

We ultimately closed about 70 percent of all roads to public travel over a several-year period beginning in 1993. This offered immediate budget relief, but we feared that a major rainstorm and flood would assault our neglected, aging road system, causing hundreds of road-associated landslides that would bleed ruin into rivers that we counted on to support salmon recovery, not to mention the corresponding loss of many millions of dollars in road asset value.

Could we develop an approach to road management that would both stabilize the roads and render them somewhat invisible in the landscape, in relation to terrestrial and aquatic impacts? We didn't trust the ability of ditches and culverts to handle storm runoff. Decades of experience had already proved their weaknesses. If neglected, they would only become greater liabilities. So before we closed the roads, we installed thousands of small, inexpensive water bars diagonally across them from the ditch side to the opposite edge of the road. We dug the water bars about a foot deep into the roadbed. These low-tech bars helped to drain all water from the road in small increments, no matter how hard it rained. Once treated, roads could be left alone. In the Coast Range, this meant rapid and dramatic changes on closed roads as grass and brush grew fast—so fast that many roads became impassible in two or three years. Toppled trees stayed where they'd fallen.

The test came in February 1996, when a tremendous storm struck the Pacific Northwest. Oregon's Coast Range endured its full-force brunt. The warm rain-on-snow of this "pineapple express" (so named because of its origins in the warm mid-Pacific) brought widespread flooding, landslides, and road and bridge losses. I will never forget flying over those mountains in a small plane when the storm abated and seeing the many surging rivers pumping a red-brown soup full of soil and trees into the deep blue Pacific Ocean.

Damage to the road system appeared to be great, as feared, and seemingly everywhere. A sad state of affairs, yes; however, we now had the chance to gauge the effectiveness of the measures we had taken over a couple of years just prior to the storm.

After the flood, we meticulously assessed the nature and extent of damage through a study of one hundred miles of road—fifty miles that had been treated with water bars and fifty miles that had received no treatment. We noted significant reductions in both frequency and severity of landslides and road failures on treated roads. Astonishingly, a detailed review showed that only one of more than 950 water bars gave way to any kind of road failure.

This was a real triumph for the "small is beautiful" school of thought. In addition, all eighty-two of the Siuslaw bridges survived the flood intact in silent tribute to the technical competence of our bridge builders. Roads now posed less risk to the health of aquatic systems.

Pressing the Priority of Fish

During my first summer on the Siuslaw in 1992, I took a trip to Mapleton to visit an innovative fish-habitat project. I left my parked car and headed for the sound of workmen in Knowles Creek, which meets the highway from Eugene to the Coast a few miles east of Mapleton, where it joins the Siuslaw River. Bill Helphinstine, Mapleton district ranger, invited me out to look at the Knowles Creek Project, promising a look at some pretty cutting-edge salmon restoration.

Knowles Creek runs only about fifteen miles from beginning to end. The creek's drainage basin includes land owned by the Forest Service, the Bureau of Land Management, and Hancock Timber Resource Group, a spinoff of the insurance giant, whose timber holdings lay nearby. Like most coastal streams, Knowles Creek had a once-abundant fishery that had been damaged by serial assaults over the years.

Huge cedar stumps dotted the floodplain, evidence of the mighty trees that had once grown there. We noticed numerous clear-cuts above us. Knowles Creek seemed typical of most coastal streams—pounded by logging, its fish largely gone, with little trace of naturalness.

It was late summer, with just a trickle of water moving. Huge logs, some as big as eight feet in diameter, lay strewn about. Helphinstine explained that the workers were trying to make a logjam. Long, stinger-like drill bits pierced holes through the logs and also into sandstone bedrock. Then wire cables were inserted through the trees and epoxied into the sandstone to anchor the logs in place.

I confess, it seemed a bit comical. Such big logs, so little water—a bit like the folly of Noah's Ark, unless you knew there was a flood coming.

Helphinstine introduced me to forest technician Lynn Hood, who enthusiastically explained how this logjam, and others like it, would slow the water and trap gravel to improve salmon habitat. I didn't see much water, and I sure didn't see any gravel.

Perhaps sensing my skepticism, Hood said I needed to come back during a winter storm, when Knowles Creek was galloping with about six feet of water. He was confident the logjam would hold tight and do its job.

I said, "What about the gravel? Where is it?"

He pointed upstream. "Oh, it's up there. Trust me. It'll come." Hood smiled, but his conviction conveyed serious belief. I'm sorry, but I wanted to laugh.

The three-way collaboration involved the Forest Service (labor and money), Hancock (logs), and Pacific Rivers Council (the design and know-how of Charlie Dewberry, the council's staff fisheries biologist). I was standing at one of about fifteen artificial logjam sites, each constructed in an effort to replicate natural conditions. Such logjams commonly occurred as a natural component of stream function when towering trees fell into the water.

Government foresters once required that timber companies remove these logs because they thought the logs obstructed salmon migration. Later, research showed that in situ wood in streams greatly improved salmon habitat. Now we were putting logs back.

Fast-forward a few years to the epic 1996 flood. The Knowles Creek vicinity was hammered. When Dewberry revisited the area, he had his hopes up. If his design was sound, the flood would have used the logjams and furthered the work that human hands had started, dramatically improving salmon habitat.

Dewberry was pleased to see that almost all his structures had held. Huge, deep gravel bars lay neatly deposited upstream from where the logjams had slowed the floodwaters. I too have been back to Knowles Creek many times. I can scarcely believe it is the same place I first saw. On my first visit, Knowles Creek resembled a sandstone bowling alley, slicked to bedrock, devoid of gravel and wood. Today it looks natural, utterly transformed.

What's more, rigorous sampling of fish numbers reveals dramatic increases. Chinook and coho salmon are now prospering where they once hung by a thread. Even chum salmon, which had completely disappeared, now return.

This is a remarkable success story. Unfortunately, salmon populations fade in literally thousands of other Knowles Creeks needing similar work. Given the saturation level of the problem, how does progress happen? One step at a time.

Another big step in our fisheries restoration effort occurred at Enchanted Valley. Located a few miles northeast of Florence, the aptly named valley is traversed by Bailey Creek, which empties into Mercer Lake and thence into Sutton Lake before meandering a couple more miles under a new name (Sutton Creek) through low, open sand dunes into the Pacific Ocean. Enchanted Valley is much like, but smaller than, many of Oregon's mid-coastal valleys due to the area's geology. Mountain streams emerge into broad, flat valleys, then into freshwater lakes, before making a brief run to the ocean.

Dairy farmers found these valleys ideal for their herds. Farmers improved their pastures by literally moving these slow, meandering streams to the edge of the valley with dikes and ditches. Farmers filled in old creek channels and removed bankside trees. This agricultural "improvement,"—usually subsidized by the USDA—nearly ruined habitat for the resident salmon. Beautiful though it was, Enchanted Valley looked nothing like the way it did in 1900. As elsewhere, Bailey Creek's salmon suffered.

Coho salmon once thrived in these systems. Adults wildly spawned in the upper forested reaches of Bailey Creek, their eggs yielding new hatchlings that would grow to become smolts in the creek's lazy meanders and the lakes' depths before making the short run to the ocean. Salmon have a strong fidelity to particular streams. Adults almost unerringly return to the same stream where they were born. Consequently, once a fishery is lost, restoration can prove difficult since native fish (unlike hatchery fish) don't often wander to repopulate neighboring streams. Numerous coho populations, including the population in the Bailey Creek area, are now endangered.

Unlike chinook salmon that hatch and return to saltwater in just a few months, coho spend over a year in freshwater. Young smolts need to survive both the low water flows and high temperatures of summer and the heavy floods of winter. Life in a ditch spells trouble, but a natural creek is ideal.

Mike Clady and Bob Metzger, both crack fisheries biologists on my staff, spoke to me passionately about the opportunity to restore Enchanted Valley. They said Bailey Creek was almost perfect for restoration. They planned on using techniques pioneered by Dave Rosgen, whose project on the East Fork San Juan River I knew of from my days in Durango.

The idea was to dig an artificially meandering channel in the middle of Enchanted Valley where Bailey Creek had once flowed, and then plug the ditch and return water to our "manufactured" creek. Reducing the gradient in this way would double the length of the creek and slow the water before it enters Mercer Lake. Importantly, such channel engineering allows the stream to leave its banks during high flows to reconnect with the flood plain. The last step is to plant shrubs and trees such as cedar and spruce to improve shading and help keep waters cold.

I grasped that, like the water bars on our forest roads, this relatively cheap concept had a high probability of success. The State of Oregon volunteered to help finance the project. If it worked, the strategy could be replicated on numerous other coastal streams to create pockets of high-value coho salmon habitat.

But once again, competing resource uses complicated what seemed at first to be a simple bonus for fish. The Rocky Mountain Elk Foundation had purchased Enchanted Valley in about 1985 and then brokered a deal transferring ownership to the Forest Service in return for assurances that the valley would be managed for elk winter range. The Siuslaw Rod and Gun Club played a prominent role in hyping this deal. The former cattle pasture became tremendous elk habitat, abetted by the Forest Service, which mowed the meadows every fall to keep the grass in a "juvenile" condition to enhance its palatability for elk. I can attest to the fact that elk loved it because I saw them on almost every visit. Hunters enjoyed having a big local elk herd that afforded high-quality hunting.

When we aired the idea of a fisheries restoration project in Enchanted Valley, controversy gathered like iron filings to a magnet.

Al Pearn, a retired military officer and active member of the Siuslaw Rod and Gun Club, lived his dream on a hillside above Enchanted Valley. Pearn, saying he represented the sentiments of the club, let me know he viewed me and my ideas as a threat. How could the Forest Service—and more particularly I, myself—even consider "ruining" perfectly good elk range for the benefit of fish? Hadn't the Forest Service acquired title to the valley by virtue of the Rocky Mountain Elk Foundation's generosity? Elk advocates strongly felt I violated the spirit of that deal. I also heard rumblings from Mapleton's staff that even they thought the proposal did not keep faith.

I saw things differently. Elk were not hurting in the Coast Range, but salmon were in deep trouble. Our actions would slightly diminish both the quantity and quality of elk habitat once the streamside trees grew to large size, but would not eliminate it. Changes would occur slowly, giving the elk herd time to adapt. If the fisheries project worked—and we thought it could, though we could not guarantee it—the benefits to salmon would be huge, and we'd have a tested, viable technique that could be applied in many other places.

At the public meeting we held in Florence, citizens had plenty to say, mostly opposing the project. They had heard that even Forest Service opinions split between elk and salmon since many Forest Service employees were avid elk hunters. Al Pearn poked his finger in my chest out in the lobby that night in Florence, saying he would "show me how democracy worked." Not surprisingly, no consensus emerged.

I ultimately decided to do the Enchanted Valley project after careful consideration of dissenting views. The Rod and Gun Club appealed my decision, but, on review, the regional forester in Portland supported my decision. We pulled money together, issued contracts, and the dirt started to fly.

The transformation of Bailey Creek was rapid and dramatic. The new streambed emerged from the construction phase, and vegetation quickly healed the raw spots. The day came to cut the water loose into the new channel, an event welcomed with nervous anticipation. Bailey Creek performed like a perfect gentleman, the water easing into the channel and slowly sauntering toward Mercer Lake. The creek embraced its new course as if returning to the haunts of its youth.

Then the fish did their job. Salmon returned to spawn, and the new channel proved a happy home.

It takes three to four years for salmon smolts to head for the ocean, mature, and return as adults to spawn. The first year those spawning adults returned to Bailey Creek, their numbers had increased by about tenfold. And succeeding years have continued to strengthen the population. Bailey Creek is now a high-quality salmon stream—and the elk continue to graze in the valley.

After I left the Siuslaw, bigger salmon restoration projects followed in several places, including Karnowski Creek and the Salmon River estuary. Johan Hogervorst, Paul Burns, Karen Bennett, and others lit things up with their passion and energy. Great things are happening. I occasionally think back to Knowles Creek and my opinion that little could be accomplished. I was so wrong.

Roll The Dice!

As environmental scrutiny became more intense and hostile throughout the 1980s, the Forest Service was losing numerous lawsuits, usually on procedural lapses rather than on the merits of a particular decision. The solution? Spend increasing amounts of time and money to armor-plate environmental documents against legal challenge. It seemed to me there must be a better way to do business.

Bernard Bormann, a Forest Service research scientist, itched to do research on our thinning program. Bormann thought we usually neglected the bigger picture when doing specific project work. Bormann's research associate, Pat Cunningham, and an editor, Martha Brookes, rounded out his research team. Owen Schmidt, an attorney with the federal Office of General Counsel, wanted to help us comply with legal requirements. Bormann and Schmidt offered their services to help us get our forest project work done smarter, cheaper, and faster.

In the Five Rivers area that fed the Alsea River, we planned to do thinning and road work over several years aimed at improving fisheries and wildlife habitat. We were committed to thinning young timber stands, but questions remained about whether our approach actually achieved the results we wanted.

Typically, the Forest Service would handle every project separately over a period of several years, the paperwork tedious, repetitious, and costly. Owen Schmidt wanted us to analyze numerous projects in one Environmental Impact Statement, and Bormann wanted to simultaneously test heavy, medium, and light thinning treatments to learn which worked best.

We stitched together the basic rudiments of our approach. We would analyze a much bigger piece of land than usual (about forty thousand acres) for a ten-year period, integrating many resource actions such as timber thinning, road closures, and fish habitat improvement. This long-term, broad-scale approach would eliminate repetitive, small-scale project analyses. Heavy, medium, and light thinning would be randomly applied in the Five Rivers drainage, along with a control where we did nothing. In some places, but not others, we'd close roads.

The simple objective: learn which strategy worked best. It may not seem radical, but this had not been done before. I had a progressive district ranger in Waldport, Doris Tai, and she had an outstanding staff to do the work.

Bormann originated the idea of testing several approaches concurrently. Each approach was designed to accomplish our objective of accelerating the achievement of old-growth forest character. The Forest Service usually analyzed a few options before selecting one preferred alternative (as required by the National Environmental Policy Act,[12] or NEPA) to implement. The remaining options were discarded and lost along with opportunities for much valuable learning, especially the value of mistakes, which can be common. By today's assessment, the entire experiment of federal forest management in the twentieth century might be viewed as an environmental mistake. Bormann asked us to be humble enough to acknowledge that our best guess might be wrong. There ought to be more than one way to do something, and testing several landscape management approaches simultaneously made sense.

Environmental Impact Statements (EISs) commonly exceeded four hundred pages and often took years to write, but Schmidt described how we could write a streamlined EIS focused on significant environmental effects. Encouraged, I asked Doris Tai and her staff to "shoot for a thirty-pager . . . and do it in four to six weeks."

In a few weeks, we met in Corvallis to review the draft EIS report, all of thirty-eight pages long. Bormann asked Pat Cunningham to roll out a big map of Five Rivers on the table. Then Cunningham explained how, working with Waldport staff, they had divided the area into twelve sectors. Now we needed to decide where our three thinning strategies and our control (leave the land as is) would be applied. There would be three replications of each.

Then they pulled out the dice. Rolling dice may not have seemed very professional, but Cunningham explained that scientific integrity required a random approach. He was serious, not budging.

Doris Tai, being the ranger, clearly had the responsibility. She just couldn't do it. I pleaded with her to pick up the dice and let destiny happen. She refused, perhaps a bit sheepishly.

I grabbed the dice and let them spill out of my hand onto the table. I savored the moment, repeatedly rolling the dice for all twelve portions of the Five Rivers area. This seemed a natural outgrowth of putting our hand to the plow and not looking back.

How different was Five Rivers from the no-holds-barred timber hey-day Forest Service I grew up in? Hugely different. Five Rivers proceeded without any appeals, as have other landscape-scale projects. The Siu-slaw continues to do more of the same at great savings. Project imple-mentation goes smoothly, and the Forest Service is better positioned to learn which landscape approach yields the best results.

Despite our good example, the regional forester's NEPA staff in Port-land remained unimpressed—opposed, in fact. They remained commit-ted to the old method of heavily armoring project-specific NEPA docu-ments designed to withstand likely appeals and litigation.

A Look Back; A Look Forward

The Forest Service arguably operated well within the mainstream of public sentiment through the 1970s. But a groundswell of opposition was mounting. A different set of societal values emerged, inviting the agency to change. The failure to respond to this value shift had profound consequences for the Forest Service. How many polls that show 90 percent of people hate clear-cuts does one have to read before concluding that it's time to do something different?

Judge Dwyer's injunction slammed the door on the past while serving as a stern reminder of consequence and accountability. But the injunction also pointed to a brighter future. Might federal forests hold the key to restoring much of what has been lost in our woods?

Many salmon populations suffered as we systematically eliminated the great forests that fostered such amazing richness. Other factors such as hatchery practices, fishing pressures, and hydropower dams compounded the insult. It is naïve to think that salmon numbers will ever approach their former abundance or that old-growth forests will ever occupy as large a portion of the landscape. But we can restore fisheries and forests to a significant degree.

Consider this. One hundred years ago, the Siuslaw National Forest was a recently burned-over young forest of seemingly little value. With time, this prodigiously productive thousand-square-mile landscape grew a phenomenal forest that exceeds ten billion dollars in timber value alone, even after previously removing several billion dollars' worth! No wonder there is still such intense debate about which values will dominate the future. But land and resources do recover. These federal forests generated many billions of dollars in profits. Aren't they now worth an investment in restoration?

Public forests—federal forestland more than state forestland—now focus strongly on wildlife and fish, clean water, recreation, and naturalness by virtue of law and public preference. Timber management philosophy is moving toward sustaining ecosystems rather than producing revenue. Restoring health and vibrancy and productivity to these lands requires work—a different sort of work than formerly known.

The Pacific Northwest, and indeed all of the United States, is blessed to have a balance of both public and private forests. Private forestlands are still primarily driven by economic forces; public forests should not be. This diversity of forest ownership is healthy because it fosters a more diverse economy. Private forests are managed more for commerce, while public forests, increasingly, are managed more environmentally.

The Forest Service emulated industrial forestry until the collapse of the big-timber era, much like a train hurtling at high speed right off the end of the track. This occurred in spite of warning signs all along the way. Yellow light—growing public contempt for clear-cutting of old-growth forests and more shrill opposition to the agency's core business. Red light—litigation tunes into the right frequency after many attempts and jams transmission. Federal courts finally stop an agency incapable of stopping itself.

In hindsight, I am still astonished that the Forest Service sold more timber in 1989 than ever in its history. Environmental groups escalated their opposition throughout the 1970s and 1980s, and the Forest Service response was to increase timber harvest. It is true that powerful political interests influenced the Forest Service timber agenda. Long-tenured western senators favored a robust timber program, and a lengthy series of Republican presidents who were traditionally pro-commerce—Nixon, Ford, Reagan, Bush—combined to weight the scales in favor of industry, and I just don't see much evidence of pushback from a resistant Forest Service.

At the local level, I felt the Siuslaw was caught in a powerful vortex of forces seemingly beyond its control. We mustered the organizational strength and courage and foresight to break out. In doing so, we discovered a hopeful future.

Since I left the Siuslaw in 1999, several forest supervisors have succeeded me in Corvallis. All chose to adhere to the Siuslaw model. Annual thinning treatments have increased to stabilize timber harvest at about forty million board feet. Ironically, the Siuslaw now sells more timber than most national forests in Oregon and Washington. Fisheries restoration abounds in key watersheds through effective collaborative efforts. The Siuslaw pushes forward with strong public support and no controversy. The environmental community now endorses the Siuslaw's embrace of forest restoration principles and pleads, "Why can't all national forests be managed this way?"

Chapter 8: Opportunity Knocks

WASHINGTON, DC, PART II, 1999–2002

Capitol Politics

Chris Wood was Forest Service Chief Michael Dombeck's policy advisor. Wood had followed Dombeck to the Forest Service from the Bureau of Land Management, where Dombeck had served, though never confirmed by the Senate, as acting director for four years. Dombeck, a fisheries biologist, had risen rapidly through the ranks of the Forest Service before detouring to the BLM, assured by then-Chief Dale Robertson that he had a "get home free" card. Now he was home again.

No Forest Service chief had ever employed a policy advisor before. It was hard to discern Wood's role exactly, but not difficult to know that he was important. I first met Wood when he came to a Forest Service leadership meeting in Oregon in about 1996. Numerous stories about his influence with Dombeck preceded his arrival. It seemed Wood served as the chief's gatekeeper. If you wanted to talk with the chief, you had to talk to Wood first.

This alarmed anyone who'd learned how to work the ropes in an agency with a very tight chain-of-command tradition. The chief's office had historically kept a very open door to field officers, especially forest supervisors. Basic respect made stopping by to say hello to the chief de rigueur when in town. Now, anecdotes dribbled through the rumor pipeline about Wood's unwelcome intervention in the affairs of the chief's office. People longed to see this arrogant badass in the flesh.

When Wood came to Oregon, I found I liked him from the get-go. He was bright, energetic, optimistic, and smiled a lot, a high-minded go-getter with a real passion for issues. Youngish, he didn't iron his shirts. He kept no pretenses, and he radiated confidence and power.

I suspect he sensed that many agency people viewed him with a wary eye, so I approached him warmly at coffee break, quickly discovering we shared a love of fishing. Then the questions started popping. He wanted to know all about the Siuslaw—what were our hot issues, what did I see as needing to happen to move Dombeck's agenda forward? I had a vague awareness Wood used such questions to gauge my land ethic. I eagerly shared our work on thinning, road closures, and watershed restoration. Based on what I'd seen of Dombeck's agenda, I thought my ideas about managing national forests were similar to the chief's.

Wood enthusiastically encouraged me to keep up the good work. He mentioned that Chief Dombeck believed we were in a time of flux, and the chief wanted to find change agents. Wood said he planned to talk with Dombeck about the Siuslaw's road closures, fisheries restoration, and thinning of plantations to promote old growth. Wood wanted me to stay in touch and keep him abreast of any new developments. This was the badass people feared? I saw nothing to fear.

To respond to Wood's curiosity, I sent him a copy of "Torrents of Change," a half-hour video produced by the Forest Service Employees for Environmental Ethics following the big flood of 1996. The video began by relating how the storm had affected the forests and rivers. It described how we on the Siuslaw had storm proofed our roads with water bars and how well our preventive measures had worked to absorb the punch of the storm. I also sent Wood a speech I'd just given at a Siuslaw forestwide celebration of human diversity. The talk included my thoughts on Forest Service obligations to tribes, since virtually all our western national forests had once been Native American ancestral homelands. Wood sent a note back saying that he really liked what he saw and would share it with the chief.

Wood must have sensed that I operated on a different wavelength than most other forest supervisors. On his next trip West, he confided to me that Dombeck had concerns about the willingness of agency leaders to break away from past traditions and dogma to assert bold change.

I agreed with Dombeck's assessment. I had a certain dread and uncertainty about the future, but a keen sense that inaction was deadly. I felt compelled to break out of the mire and do something.

As we charged ahead with our programs on the Siuslaw, I mentioned to Wood how much I appreciated the newfound support I had received from the environmental community. He could appreciate that because it was so rare. Also, a fresh energy and enthusiasm was evident within the Siuslaw organization as we pursued our new mission. Yet, I explained, many of my peers had little use for or interest in the Siuslaw's resurgent agenda.

I remember that Wood said something like, "We need this kind of stuff back in DC."

On Wood's next trip, he came right out and asked if I'd consider coming back to work in the chief's office. "As what, for instance?" I replied. I loved my work in Oregon. Besides, I'd already spent a two-and-a-half-year tour in DC. I was not eager to return. In fact, going back to DC had not occurred to me since I arrived in Oregon.

"Well, the associate chief and deputy chief for national forests are both open."

I let this cannonball bang in my brain for a moment. No forest supervisor had ever made that kind of jump.

I found Wood's suggestion extraordinary. The associate and deputy positions were, respectively, the number-two and number-three positions in the agency. Both fell under the category of Senior Executive Service (SES) appointments, which normally involved an arduous evaluation just to get qualified, then usually six months of "charm school" training. I had done neither. Furthermore, plum positions like a deputy chief's post always went to people already in the queue for SES positions, like regional foresters or associate deputy chiefs. This was how the old Forest Service did things.

Undeterred, Wood said the chief didn't put much stock in all that. Dombeck just wanted to get people in these jobs who would move the Forest Service ahead. Wood said they had ways to cut through all the bureaucracy—evidenced by Dombeck's recent appointment of

Francis Pandolfi, an executive from the Times-Mirror Corporation and a complete outsider, to an SES job through open competition. Dombeck wanted people with new ideas, big ideas.

I asked if Dombeck was serious. Wood said yes, but he could guarantee nothing, since both positions would be filled through open competition. Dombeck wanted someone like me, but he couldn't hand me the job. I understood.

I told Wood I needed time to consider his request.

I knew something about the recent rapid changes at the top. Almost immediately after Dombeck became chief, he'd fired several top SES leaders, including the former deputy chief, Gray Reynolds. Reynolds had issued a bitter rejoinder that had hit the inbox of almost all employees, saying his firing was unwarranted, unwelcome, and an offense to Forest Service tradition. Dombeck had replaced Reynolds with Bob Joslin, southern regional forester from Atlanta, but Joslin retired after only a year and a half amid rumors that he couldn't handle the environmental politics of the Clinton administration.

Gloria Manning stepped in behind Joslin as the acting deputy chief.

There was turmoil elsewhere. Phil Janik had replaced Dave Unger, the recently retired associate chief. Dombeck chose to split Janik's duties to create a second associate chief position. The person filling this new job would supervise the deputy chief of the national forest system.

Substantive questions needed answers before I could even think about being that person.

Could any forest supervisor succeed in such an environment, especially a supervisor from the Siuslaw, which was a small forest relative to many others? I would not be regarded as a particularly powerful or credible forest supervisor. I anticipated strong resistance and pushback because of an agency culture that respected power. If Dombeck selected me, largely unknown, as either associate or deputy chief, I would leapfrog about four or five rungs on the organizational ladder. How would this affect agency leaders' trust in the chief? Bypassing all the more likely candidates would send a strong signal that he doubted them. This would echo in their doubts about him.

Could I succeed? Regardless of how positive Dombeck was about my accomplishments, he didn't really know me, and I didn't know him. My biggest concern was not my ideology, but my capacity to develop and build followership. Would regional foresters and national program directors willingly roll up their sleeves with me to get things done? A long career in the Forest Service had convinced me that a failure in either of these important jobs would be costly to both the chief and the agency.

I harbored doubts. Such a move carried steep risks. If Dombeck selected a nontraditional person like me, it would be difficult for me to bring others along, and I could easily become isolated. Yet I also sensed that Dombeck intended to pursue an ambitious agenda, and he wanted and needed like-minded people to help him. The possibility of helping shift the trajectory of the Forest Service held a tantalizing appeal.

I had long talks with my wife, Judy. Surprisingly, she supported my going after the job. She knew the inherent uncertainty that I'd get either job. Yet she'd enjoyed living in the DC area. We both knew a promotion would bring a big jump in pay, too.

Until now, Chris Wood had done all the talking with me. I told Wood that I thought it was time for me and Mike Dombeck to speak directly. Could Dombeck call me personally? Days went by, and then weeks. I began to wonder whether all my speculation was for naught.

Then, as often happens, about the time my anxiety began to burn away, Dombeck called.

We had a good talk. I shared my concerns and asked the chief his thoughts. He said he admired my accomplishments on the Siuslaw. He thought I had what he and the Forest Service needed to move the agency forward. He mentioned the scarcity of big ideas, much less big accomplishments, in crafting a twenty-first-century agenda. He and Jim Lyons, the under secretary of agriculture, both wanted people on the leadership team who would forge change.

We discussed the SES hiring process. He said he had hired others without SES experience via open competition. It was a tricky deal, though, and I needed to quickly get acquainted with Tom Leeper, who was the Forest Service's SES expert. Dombeck asked me to tell Leeper, "The chief wants you to shepherd me through the application process."

It was now summer 1998. Wood had mentioned that Sally Collins, Deschutes National Forest supervisor in Bend, Oregon, was also in the running. Collins and I tended to think alike, and we got along well. She and I discussed the merits of going after these jobs, and we both thought it was a big gamble. Collins told me her daughter was entering her senior high school year in Bend, and a move at this time seemed imprudent for the family. I countered that it would take many months before a decision was made, and she could bargain to delay her starting date until after graduation the following June. After much soul-searching, Collins decided not to go after either job.

I called Tom Leeper. He stood ready to help, but he warned of the arduous process before me. To receive a favorable review by the Office of Personnel Management, I would need to put together a sterling application. He said he would coach me along the way. I did ask if being a GS-14 would be held against me, since almost all those who qualified for SES selections came from the higher-level GS-15 pool. He said that would not be a problem.

The process required that candidates develop narratives demonstrating their competence in five evaluation factors: leading people, leading change, building coalitions, being results-driven, and having business acumen. Applications had to be approved by a panel of three government reviewers. If rejected, the application could be revised and submitted again, but only once.

I dove in and spent many evenings preparing my application until I felt it represented my best effort. I ran it by Tom Leeper, who said, "Looks OK." So I submitted it. Weeks went by before Leeper called to tell me I hadn't made it. The panel found four out of my five elements "deficient," and the fifth just marginal. Leeper said I had another shot, but I had to pass the second time or it would be a dead end.

I begged for some insight on what I had done wrong and how to fix it. Leeper said they really looked for a "certain style" of writing (what I privately described to myself as elite bureaucrat or self-aggrandizing). Could he provide some examples? No, but he'd work with me when I sent him the next write-up. This he did: he said the application was still not there.

Desperate now, I asked if he could rewrite a few paragraphs to illustrate what he meant by a certain style. Then I submitted another heavily edited round for the Leeper test. Leeper remained nervous about my application, and I even more so.

I then invoked the Dombeck imperative. To "shepherd me" through this process, I told Leeper that he needed to give me more hands-on help. I told him that Dombeck wanted me to be a candidate available for selection. If I didn't get through the SES gate, we'd all look bad. I asked him to work with me line by line, editing it strongly to make sure it passed the SES panel review. I was grateful when he helped me prepare my last-chance version for reconsideration.

I later discovered a couple of interesting things. Francis Pandolfi confided to me that he applied for his SES job with just one handwritten paragraph for each of the five evaluation elements. Seeing this, Leeper had delicately advised Pandolfi that this wouldn't do, since SES applications were serious business, and Pandolfi had a lot of work to do before he could submit his work. Pandolfi said he told Leeper, "No, Tom, you've got a lot of work to do." Pandolfi got the job after Leeper polished those few paragraphs to a luster that passed the SES board.

I talked with another SES person who served on review panels. He said the Office of Personnel Management took a very dim view of GS-14 candidates (like me), deeming them unworthy of consideration, since SES candidates were usually from the higher-level GS-15 pool. He indicated that GS-14s were routinely rejected without serious consideration.

Do these anecdotes indicate a double standard, or was I simply naïve? Even though I think Leeper should have been more helpful from the beginning, my application squeaked through the second time, and I became an applicant in good standing for both high-level jobs. As associate chief, I would report directly to Dombeck, and my responsibilities would include national forests and the agency's research branch. As deputy chief, I would report to the associate chief, and I would manage all national forest issues, supervising nine regional foresters, ten national program directors, and a staff of about fifteen, with an overall budget of about two billion dollars. I knew I would be thrilled, and fully challenged, if offered either one.

The Gauntlet

After I knew I was a contender for two of the top jobs in the Forest Service, I felt I'd better do some background research on the top-level culture of this agency I'd spent my career in. I selected three people to confer with.

Everett Towle, planning director for my first tour in DC, had earned my respect as a wise person. Towle was retired and living in his dream home in southern Maine. When I visited him there, I asked if becoming either associate or deputy chief would be acceptable and workable based on his assessment of agency culture. Towle said there was good news and bad. The good news was that forest supervisors, as a group, were highly respected within the agency culture. The bad news was that many would still view my appointment with cynicism because it had not followed the normal promotion procedures. It would be tough sledding.

Dave Unger, a retired associate deputy chief and Ev Towle's former boss, came to the Forest Service from a high-level position in the Soil Conservation Service. He would know something about stepping into a key position without internal support. Unger told me it would be difficult, but if I understood this and had a strong agenda to pursue, it would be worth it.

Unger asked me, "Do you really want the job?" It was a fair question. I'd be unlikely to succeed unless I hungered for the responsibility and the challenges that came with it.

The third person I consulted was Andy Stahl, executive director of FSEEE in Eugene, Oregon. It's fair to say that Stahl was generally despised by the Forest Service leadership. He and FSEEE existed primarily as goads to the agency. Stahl enjoyed a good fight and did not hesitate to start one when he thought the Forest Service erred. We'd worked together enough that I knew I could count on straight talk.

I told Stahl, "I'm pleased that Dombeck is looking at me, but I think these jobs can be pretty nasty and thankless." Stahl said I would need to have an iron backbone, learn how to say no, and be willing to absorb insults without the need for revenge. He suggested I craft a short list of personal priorities that I intended to achieve to avoid the pitfall of being very busy with the trappings of responsibility but accomplishing nothing.

I asked for confidentiality and thanked each for their counsel. I knew either job would be a difficult challenge, but also the chance of a lifetime. By this time, my preference had gravitated to the deputy chief job, which I thought fit my skills better than the associate chief position. I sensed Dombeck was a new kind of Forest Service chief, committed to putting the agency on a new track. I liked where the chief was headed. I wanted to be part of his team.

I also knew that a presidential election was scheduled in November 2000. If we elected a Republican president, Dombeck would certainly quit or be fired. Though I would not be a political appointee, I'd probably be ushered out the door right behind him whether I wanted to go or not. On the plus side, I'd had a long and full career and I could stand that outcome. This was a risk worth taking.

The waiting was agony. Chris Wood took on the role of mole, giving me occasional updates. He assured me that things were moving toward my selection for one or the other job. "Be patient, Jim," he counseled.

Bat Box

About March 1999, Stahl, who occasionally represented employees who thought they were getting squashed, called me about a case involving Craig Cope, a Bighorn National Forest wilderness manager. Stahl knew I'd worked on the Bighorn. Did I know Gail Kimbell, the current Bighorn forest supervisor? I said I'd never met her, though I once tried to hire her for a job on the San Juan. I told Stahl I thought she had a generally good reputation.

Stahl said Kimbell planned to hammer Cope for something petty. The Forest Service had mounted a roosting box for bats on a power pole, putting it right smack in the middle of the paved office parking lot. Cope thought bats would not roost there (probably true), so one Saturday he decided to move it to the edge of the lot next to some trees, using government tools: post-hole digger, shovel, and such. This made the district ranger furious. She convinced Kimbell to propose a three-week suspension without pay, a very serious penalty. About the only thing harsher was getting fired. Stahl planned a full-court press to publicly embarrass Kimbell and the Forest Service as best he could, a task he excelled at. He promised to back off if I could coax her to reconsider.

I called her. She seemed irritated about my butting in on her business. True, I was. I explained that Stahl loved a good cause. The Bat Box had a nice ring as a notorious cause. No doubt Cope would be sensationalized in the media as a helpless victim. No matter what penalty he deserved, the Forest Service would come out the loser, painted as an unreasonable, stuffy tyrant. I didn't tell her what to do, but I asked her to just please consider the bigger picture.

Almost immediately, Chris Wood was on the phone. Kimbell had called Lyle Laverty, the Rocky Mountain regional forester, who then immediately called the chief to tell Dombeck he needed to get Stahl's "attack dog" on the leash. That would be me. Wood, agitated, said my selection as deputy chief was at a critical juncture—stuff like this could imperil my appointment.

But worse yet, Laverty had called Stahl first. Stahl couldn't resist smugly telling him that he better be careful who he messed with, because I was "going to be his boss." I was dismayed. Stahl was way out of bounds, and he had violated my request for confidentiality.

Laverty freaked out, giving Stahl exactly the reaction he hoped for. Then he called Dombeck again to ask if it was true and pepper him with choice words about how come Stahl knew all this. This incident set the tone for a very frosty tenure as Laverty's boss.

Gail Kimbell imposed the suspension, Andy Stahl and FSEEE got their pound of flesh in the media, and Cope later joined with several Bighorn employees in a complaint against the Forest Service, naming Gail Kimbell, Lyle Laverty, and Tom Thompson, deputy regional forester, as principals. A judgment against the Forest Service awarded Cope and the others several hundred thousand dollars.

I was disappointed at the pettiness of these individuals handling the Cope case because I thought it could have been easily resolved. This reminded me of the same defensive mentality that seemed to drive agency officials when it came to resolving disputes with environmentalists.

A Weird Welcome

After months of tense waiting, I got the news I'd been waiting for. Mike Dombeck announced that I was the next deputy chief of the National Forest System (NFS).

Like many big events in life, it seemed surreal. The chief invited me in from Oregon for a "meet and greet" with lots of folks. The public affairs staff took pictures and prepared a formal press release. The day slipped away in a blur of activity, all with a rich, celebratory feel.

Dombeck walked me down to his weekly executive session. I was not well-known to his leadership group, though I had worked in DC before. I'd once met Gloria Manning, now the acting deputy chief for the NFS. She'd assumed the position when Bob Joslin retired many months ago. The job had taken a long time to fill.

Manning and I each worked for Ev Towle in the planning shop in the late 1980s. Manning had moved on just before I arrived, later serving as deputy regional forester for Bob Joslin in Atlanta. Joslin subsequently brought her to DC as his associate deputy chief. After my appointment, I expected her to return to her old associate deputy chief position, where she would report to me.

After the pleasantries, Dombeck conducted a brief executive meeting, and then we adjourned. As we left, Manning said while in the hallway, "I know you'll probably want to replace me. I'm ready to go. And by the way, I haven't had any time off in a long while. I'm tired. So I'm taking a six-week vacation."

I was stunned. Groping for words, I said something about that being presumptuous and premature, and that I had no intention of immediately replacing her. We'll need to talk more about that later, I said.

It troubled me that she would leave for six weeks when a solid transition would be helpful to me. I later learned that she had not applied for the deputy chief position. She didn't want it. I also learned that Manning expected to become a regional forester after departing my staff, preferably in Portland. It seemed a bizarre, weird welcome and not entirely out of character with the whole experience.

Chapter 9: Roadless Area Climax

WASHINGTON, DC, PART III, 1999–2002

I chose as one of my first tasks as deputy chief to review my short list of priorities, which I'd prepared in accord with Andy Stahl's suggestion to sustain my focus. My highest priority was to move the Forest Service toward stewardship contracting rather than timber sales as its primary land-management tool.

The agency had developed its traditional tool of choice—the timber sale contract—as logging ramped up in the 1950s. Although the timber contract had been modified many times over succeeding decades, it remained stuck in the rut of selling stumpage (standing trees) to the highest bidder, a generally acceptable method except in monopoly markets or when various buyers engaged in collusion.

The contract did contain an effective mechanism for loggers to trade timber value for road construction. That was a valuable provision during timber's heyday, but national forests had little need for new roads. What the Forest Service desperately needed now was a modern, less costly mechanism to manage forests for restoration and healthy landscapes. Stewardship contracting held enormous potential to improve our way of doing business.

I learned that Mike Dombeck also had a short list that subsumed mine. I knew of his focus on watershed restoration. In fact, my interest and accomplishments in this area had everything to do with my being selected as deputy chief.

On another front, Jim Lyons, under secretary of agriculture, convened a committee of scientists headed by Norm Johnson, an Oregon State University forestry professor, to lay the groundwork for a new National Forest Management Act planning regulation to replace the 1982 version. Johnson's lengthy involvement with forest planning, beginning with his developing the FORPLAN computer model, provided good perspective and insight to lead the comprehensive review of NFMA. At Lyons's request, I served with several other agency staff on an advisory group helping a Forest Service technical team develop this new regulation. This important undertaking would affect every national forest.

But a major surprise lay in my path. Within days of my assuming office, Dombeck informed me that the Forest Service would undertake an additional regulation to protect roadless areas. Although I had a hint of this proposal from an earlier moratorium on road construction within inventoried roadless areas, the audacity and sweep of this proposal stunned me.

For decades, roadless areas had been the source, or pool of lands, from which Congress designated wilderness. Roadless lands also served as the best source of new timber because they generally had little or no prior logging. Roadless areas enjoyed no official status until the Forest Service undertook a national Roadless Area Review and Evaluation[13], begun in 1967 and completed in 1972. The RARE review, as it was called, recommended that 12.3 million acres be set aside for wilderness. However, the accompanying environmental impact statement was challenged in court and judged to be deficient.

USDA's assistant secretary Rupert Cutler, a former executive with The Wilderness Society, initiated a subsequent Roadless Area Review and Evaluation (RARE II)[14] during the Carter presidency in 1977, seeking to improve on the original study. RARE II recommended 15 million acres for wilderness and another 10.8 million for further study.

Given this heightened interest, subsequent Forest Service logging and road-construction activities in roadless areas were often stopped by environmental lawsuits. The environmental community attached great importance to protecting the wildness of roadless areas, and the environmental community regarded the Forest Service as a threat to protecting roadless-area values.

This values conflict echoed the lengthy battles over cutting old-growth forests. Forest Service dogma called for logging to convert old forest to "healthy" stands of young, vibrant trees. Yet the agency's own research increasingly revealed that old-growth forests contained incredible diversity and complexity and served vital ecosystem functions. Far from being worthless, old growth held astounding value if one looked beyond mere timber products. Jeff DeBonis's 1989 letter to Chief Robertson had been the point of the spear in halting the systematic elimination of old-growth forests.

Similarly, roadless areas had long been regarded by the Forest Service as merely the next place one goes to log more trees. Even so, in 2000, after nearly one hundred years of Forest Service management, there still remained about 58 million acres of roadless areas—about 2 percent of the land area of the United States and equivalent to an area the size of Wyoming. Congress had already created about 36 million acres of wilderness. Thus, in aggregate, about half of the 192 million total acres of national forest remained undeveloped.

Wilderness areas were predominantly high-elevation lands. Roadless areas, in contrast, occupied much more of the niche of mid-elevation lands that tended to be more heavily forested. The timber industry, as well as many people in Forest Service leadership, continued to view roadless-area protection as antithetical to multiple-use mandates. The Forest Service seldom championed the wilderness cause, and could even be said to resist it.

For example, in 1984, after a crippling recession had hurt companies with Forest Service timber contracts, Oregon's Senator Mark Hatfield steered a bill through Congress that traded wilderness protection for a "release" of some roadless areas for immediate timber harvest. I recalled, fresh upon arriving on the Siuslaw, looking at a road leading right into the bulls-eye of what had been one of the Siuslaw's few remaining large roadless areas near Hebo, Oregon. When built, the road (which accessed several 1985-era clear-cuts) permanently nullified the area's unmanaged, untrammeled character.

These roads were legal, thanks to Hatfield's legislation. However, the "release" language did not require that roadless lands be compromised by logging and road construction. The swift action taken by the Forest Service eloquently symbolized the agency's operational values. Protect roadless areas? Not if there were logs to be had. Furthermore, acting to compromise the area's roadless character eliminated the nuisance factor, making future management less complex.

Roadless-area protection and timber production had been engaged in a slow-motion collision for at least twenty years. It seemed clear to me that the Forest Service controlled both variables. Thus, the Forest Service made itself the greatest threat to roadless areas. When Chief Dombeck proposed that the Forest Service develop a regulation for permanent protection of roadless areas, agency traditionalists felt downright insulted.

But Chris Wood, Dombeck's policy advisor, said that, with roadless-area protection, "The Forest Service is becoming relevant again." I won't forget that remark. The sharp contrast between insulting and relevant made me ponder. I took Wood to mean that roadless areas were a huge issue to a vast constituency of environmentalists who viewed the Forest Service as an abuser rather than a protector of roadless areas and public lands in general. I thought this perception should have greatly alarmed agency leaders, but it didn't seem to.

Dinosaurs?

The fissures that appeared after Dombeck's proposal revealed the agency as something less than one big happy family. Gloria Flora, Humboldt-Toiyabe National Forest supervisor in Nevada, made a rather public offhand remark about "dinosaurs" in Forest Service leadership who needed to retire. This outraged many forest supervisors who demanded that Chief Dombeck disavow her remarks.

Dombeck called me on the phone. "Could you come see me right away?" I took the back stairs to his office, which was just above mine. He asked me to read a letter he'd just received from Mike Lunn, Oregon's

Rogue River-Siskiyou National Forest supervisor. Dombeck told me the letter had been copied to "almost everybody" in the Forest Service. He asked what I thought he should do.

The letter resembled the DeBonis letter in a fashion. It was a direct, personal appeal to the chief outside the normal chain of command—except that it spoke to agency culture, not land management. Mike Lunn took umbrage at Flora's statement, and he gave the chief a hard-edged testimonial as to how the "dinosaurs," including himself, had built and sustained a great organization. The letter was a forceful defense of the old guard, tinged with defiance, contempt, and disrespect. The letter was also honest. Lunn was clearly calling Dombeck out as to how the chief felt about dinosaurs. The chief had an upcoming interview with *Time* magazine and now expected questions on this brouhaha.

I remembered Chief Robertson's bungling of the DeBonis letter. That had left me with an important lesson to be applied here and now: Lunn's letter demanded an immediate response.

I first suggested that Dombeck call Lunn to thank him for his letter, for having the guts to tell him what he was thinking. I told him, don't argue; just be open, and listen if he's got more to say. Then reassure him that you value agency leaders and all they do and have done for the Forest Service. I said that Lunn needs to hear that the chief of the Forest Service doesn't consider him and his colleagues to be dinosaurs. Then, I said, write him a brief response, reiterating these few points. Address the reply to Lunn only. If Lunn wants to mail it around, let him. Don't escalate, but don't be dismissive.

Dombeck did this. In a few days, many thousands of agency employees had seen Lunn's letter as it spread via email. In these days before blogging, hundreds of commentators piled on via email using a "reply all" command, so that the effect multiplied geometrically. Folks generally piled on adulation for Lunn's "telling it like it is!" I recall almost no recognition of Dombeck's response.

Emboldened, Lunn wrote yet another letter, basically doubling down and getting in more licks. Dombeck took the high road—no response this time. But the Lunn letters boldly revealed in our field leadership a lack of trust in Dombeck and feelings of betrayal. Hard stuff.

Ticking Clock

Hermann Gucinski, a Forest Service research scientist in Corvallis, was lead author of Forest Roads: A Synthesis of Scientific Information.[15] This comprehensive report described both the benefits and the adverse consequences of roads and road use. Other research demonstrated that roadless lands are the best landscapes in terms of water quality, fish and wildlife populations, endangered species, and healthy forests. These lands are also bulwarks against invasive species, because they have an essentially robust natural character that road construction tends to degrade.

The Forest Service now found itself caught in a powerful vortex, a result of failing at the critical task of solving prominent problems. Serving the public demanded that the Forest Service craft a progressive policy to deal with roadless areas and resolve the issue.

I found it remarkable that politicians in the Clinton administration trusted the Forest Service for the undertaking. One overwhelming, simple reason accounted for that: Chief Mike Dombeck. The chief often lamented that the Forest Service had become extremely defensive and apprehensive trying to defend its logging legacy. "It's going to be a lot more fun playing offense," he'd say, grinning broadly. He believed roadless-area protection would be very popular and likely to win the admiration and restore the trust of a cynical environmental community. News media often editorialized about how Forest Service leadership drifted weakly in the face of public antipathy, seeming to have run off the rails. Dombeck relished the idea of sitting down in front of editorial boards from the *New York Times*, *Washington Post*, and *Los Angeles Times* to discuss the opportunity to contend for roadless-area protection.

Several difficult realities confronted the Forest Service in resolving one of the most sweeping and controversial issues of the past several decades. Completing a regulation governing fifty-eight million acres would affect virtually every national forest from coast to coast. Opposition would quickly mount, particularly from the conservative western states that held the vast majority of roadless acres. As anticipated, but sadly, some of the most intense opposition would come from within the Forest Service.

And there was the clock. This was summer 1999. A presidential election loomed in November 2000. Political reality required that a regulation be completed before President Clinton left office. There were two reasons: first, the Democrat Clinton wanted credit for a key environmental accomplishment; second, we didn't want to waste the momentum, because we knew the election of a Republican would stop it dead. Tick, tock, tick, tock... We submitted a project calendar with all the requisite time periods—sixty days for this, ninety days for that—indicating we could complete the project by mid-December 2000, after the election and before the January inauguration of a new president. An achievable assignment, but one with no margin for error or delay.

The new NFMA planning regulation was under way at the same time. This was a very different but equally weighty undertaking that promised to seriously complicate the roadless protection effort. Under secretary Lyons viewed the NFMA rule as a high priority; it was his passion. The roadless rule was Dombeck's.

The Clinton administration demanded that the Forest Service complete both regulations during this presidency. In tandem, the twin endeavors of planning and roadless regulations were a huge burden that seriously taxed the capacity of the agency.

The associate chief position remained vacant. Shortly after I arrived, we interviewed candidates, only a few, with almost none from the Forest Service. Dombeck ultimately selected Hilda Diaz-Soltero, a National Marine Fisheries Service executive who also had extensive experience with the Fish and Wildlife Service, again sending a signal that he was shaking things up.

Diaz-Soltero and I became joined at the hip over the next couple of years as the Forest Service endeavored to complete both the NFMA planning and roadless regulations. She had passion and fire. Without her unstinting efforts, I'm not certain either regulation would have been completed.

We already had a team and structure in place to craft the NFMA planning regulation. It needed the usual oversight and encouragement.

But we had no organization for the roadless issue. We were figuratively entering unknown territory without a map. Dombeck approached me with the challenge to find someone to lead the roadless effort—one of the most significant Forest Service challenges in its history. The job needed to be done well, it had to be done on time, and failure was not an option. I thought to myself, "I've got just the person in mind."

Bill Sexton, my old boss on the San Juan, was now assistant director for the Forest Service's Ecosystem Management Coordination office in DC. He had openly voiced his hopes to become a regional forester. I mentally connected the dots—take on the roadless issue, get it done, and the Forest Service will owe Bill Sexton a favor. Name your dream job. It's yours! It didn't hurt that Sexton was capable of doing just about anything he put his mind to.

I spoke with Sexton and made the pitch. His response was matter-of-fact: "Not interested. There will be a Republican president in 2000. Anyone who touches this roadless thing will be dead meat."

I bristled. I had not expected such a blunt assessment, with its unstated implication for my mission. So much for nuance. In that moment, any thought I had of cajoling or persuading Sexton into leading the roadless team vaporized. He had made a valid point about the political realities, but it was essential to have a leader who believed in the work without reservations. I think some things are so consequential that personal considerations should be waived. When my wife and I discussed the circumstances of my becoming deputy chief, we both assumed that if a Republican were elected I'd be out a job. I took the job anyway. I voted for Bush anyway.

I called Sharon Heywood, Shasta-Trinity forest supervisor in California. Not interested. I put out an SOS call to all regional foresters. I needed help, urgently. Regional foresters seemed stumped. I think they agreed with Bill Sexton's sentiments. Why sacrifice anyone?

Finally, Brad Powell, regional forester for California, called to say he had someone in mind: "Scott Conroy on the Modoc."

I felt a deep anxiety. I had met and knew most of the 120-plus forest supervisors, but I'd never even heard of Scott Conroy. And few considered the Modoc National Forest (office in Alturas, in the far northeast corner of California) to be an impressive locale. It seemed we were nearing the bottom of our candidate pool. I reminded Powell that this was the most important thing the agency had done in decades. Can we trust this issue to Conroy? A lot hangs on the outcome.

Powell told me that he thought Conroy, though a sleeper, might be just what we needed. He was competent, very disciplined, and could be relied on to get things done.

I reported to Dombeck on my efforts to find a person to lead the roadless effort. I noted my disappointment that the regional foresters couldn't help us identify any top-flight forest supervisors. I also mentioned the Bill Sexton episode. Dombeck noted that at least Sexton was honest and agreed that we didn't want him: "We need a believer."

I got the nod to vet Conroy. Frankly, I had few options left. And it occurred to me that the reason I was now deputy chief was because Dombeck chose to trust me. Maybe I needed to trust in Conroy's competence simply because the Forest Service always tried to pick capable people for forest supervisor jobs. That, coupled with Powell's endorsement, encouraged me. Then everyone I spoke to enthusiastically supported Conroy. Now it was time to speak to him myself.

Conroy immediately impressed me with his deep desire to help the Forest Service accomplish this historic task. He knew it would be tough sledding, but he liked a challenge. My concerns that he might be unequal to the task, or worse, oblivious to the difficulties, waned. Importantly, he believed roadless areas needed protection and wanted to be a part of it.

I said there were no guarantees, but if we could pull this off, he'd likely get a good promotion. Was he interested in greater responsibilities? Yes!

We had our team leader. Conroy and his wife made immediate plans to move to DC. I called Brad Powell to thank him for his help, though I still harbored doubts that we could succeed.

Making Hay

In the next few months we seriously ramped up the roadless protection effort. We published a Notice of Intent in the Federal Register to prepare an Environmental Impact Statement and regulations to achieve roadless-area protection. Scott Conroy immediately began recruiting his team. We swung into our public involvement effort, which became the most ambitious in agency history. We conducted roadless-issue meetings for all 125 national forests, as well as many formal public hearings. We cooperated with other agencies such as BLM, Park Service, and Department of Justice. It was an incredibly hectic full-on press.

Industry groups and western Republican governors whipped up strong headwinds to blow the effort off course or, better yet, onto the rocks. Since most of the roadless areas were in the West, Chris Wood and I traveled to Salt Lake City to meet with several governors' representatives, all suspicious of the political overtones of the effort. The meeting was hostile and tense.

"Why now?" was the trenchant question. Everybody recognized roadless protection as a potent issue. We all knew the clock was ticking. Wood gamely explained that roadless issues had grown and festered for decades. It served no one's interest to leave this important issue unaddressed. While Wood spoke, I got the feeling the gang was searching for the best way to stop us.

Wood and I had anticipated that some of them would ask for cooperating agency status, which would essentially make them partners in our effort. This sort of collaboration is allowed throughout the process of writing an Environmental Impact Statement under the rules of the National Environmental Policy Act. But this would effectively give them power to sabotage the project. And they did ask, but we fended them off, saying that it would be impractical and unworkable, since so many states were involved. If we made one state a cooperating agency, we'd have to do the same for all. We vowed to keep the states' governors informed throughout, but we were firm that only federal agencies would be cooperating agencies.

Pandemonium ensued. When Montana's representative pressed us hard on this issue, we asked why he felt this privilege was essential. He said, "We want to delay the process so that you can't finish the job before Clinton leaves office." They would take their chances from there on a hoped-for Republican White House pulling the plug on roadless-area protection.

Silently, I gave him high marks for honesty. After overcoming my shock at his brazen position, I stated that this was not a valid rationale for a "cooperating" agency. We left with a firm "No. You'll be hearing from us. Thanks for coming."

Adversarial states were not our only obstacle. While we received great cooperation from other federal agencies and resolute assistance from the Department of Justice, difficulties began to develop with the Office of Management and Budget, whose Office of Information and Regulatory Affairs (OIRA) evaluated all new federal regulations for their financial implications. All our work traveled through OIRA for review and approval at every step, and the inevitable rewrites necessitated seemingly interminable delays.

Normal operating procedure consisted of OIRA's analysts reviewing our work product, then transmitting questions back to us for a response. In our case, we might get fifty to sixty questions in a five-page memorandum, iron out issues, and move ahead. Repeat. But under the duress of a compressed time schedule, we had no time for "normal."

OIRA usually gave us a deadline of a week or two for our response. But OIRA had no deadlines for responding back and often responded by giving us more questions about our answers to their original questions. In my view, OIRA's questions betrayed a fundamental ignorance of the Forest Service, sometimes along the lines of, "Just what does the Forest Service do exactly?"

I finally had a heart-to-heart with the supervisor of our OIRA analyst. I expressed my frustration with the lack of pace and focus. The White House, I told him, had "expectations." By that I meant we both knew heads would roll if we didn't get the regulation completed on time. Could we agree that OIRA needed to turn up the burner to ensure that we got the job done on time?

"Not really," he replied. I could scarcely believe what I was hearing. He informed me that OIRA had its way of doing things, and he saw no need for adjustments. He understood my concern, but the Forest Service was really no different than any other agency in need of its services. In fact, he noted that because the analyst was uncomfortable being in the same room with me, OIRA would appreciate it if, henceforth, I would absent myself from further interactions. "Please send someone else to meetings with OIRA."

I was floored. I don't consider myself a bully. But simple pragmatism required that I swallow my pride and ask Scott Conroy to handle future meetings with OIRA. Remarkably, we worked through difficulties, though OIRA remained a nettlesome impediment. Conroy was good! Better than me.

Other aspects of developing the rule flowed well. We held hundreds of public meetings. The volume of comment on the rule was massive, unprecedented. With Pew Environment Group funding a big advocacy campaign, the Forest Service received well over a million comments on our draft proposal. We had no precedent for processing and analyzing this volume of feedback, yet a task team heroically accomplished this as well. The trajectory of our effort held pretty tightly to the project calendar. We were still optimistic about concluding our work in December 2000.

Alaska, as usual, required special handling. Among the fifty-eight million acres of roadless areas, Alaska's two national forests, the Tongass and Chugach, contained fully 20 percent, about twelve million acres. Of the two, the Tongass National Forest in southeast Alaska (remarkably, a bit larger than West Virginia) has long been a political hotbed. With eleven thousand miles of coast and fifty-seven thousand miles of streams, the spectacular Tongass is regarded by the environmental community as the crown jewel of all national forests. For its part, Alaska's congressional delegation historically regarded the Tongass as its parochial domain, passing out logging and mining favors to local residents and businesses. The Forest Service is caught in a pincerlike grip between environmental constituents and powerful politicians.

With 9.4 million acres of Tongass roadless area distributed throughout this vast 17-million-acre archipelago of coastal rainforest (along with 6.5 million acres of designated wilderness and National Monument lands), the destiny of these roadless areas loomed as a huge unresolved issue. Rather than propose any particular preference in the draft Environmental Impact Statement, we chose to defer the ultimate fate of Alaska's roadless lands to the final decision phase and simply invited public comments about the issue.

Mike Dombeck asked my opinion about the best future for Alaska's roadless lands. By this time I had hired Sally Collins, my friend from the Deschutes National Forest, to be an associate deputy chief. Having never visited Alaska, I told Dombeck I lacked a good enough feel for Alaska to provide a sensible answer. Thus, in August 2000, Collins and I made an unforgettable trip to Alaska with the chief's question in mind.

We immersed ourselves in the splendors of Alaska, accompanied throughout by Forest Service officials, primarily Rick Cables, the regional forester. Alaska's enormous spaces, raw exuberance, and uniqueness left me deeply touched, almost speechless. All Americans should visit Alaska to let it speak to their souls.

Over a period of several days, Cables sought to convey to Collins and me that Alaska's roadless lands needed to remain available to local citizens for subsistence, commerce, and the basics of community sustainability. Uniquely, almost all of southeast Alaska is national forest land, meaning Forest Service management and decisions govern every aspect of life for Alaskans. In brief, Cables believed the lands' greatest value lay in utility, not protection.

Upon our return to Washington, Collins and I met with Dombeck to discuss our views on protecting Alaska's roadless lands. We described as best we could the nature and complexity of Alaska's roadless issue, ultimately siding with Cables's view that comprehensive, extensive roadless protection was unnecessary. Dombeck's face remained deadpan and he said little beyond thanking us. The few questions he did ask seemed rhetorical. I left with a sense that he disagreed with and was disappointed in our position.

Ultimately, the Forest Service decided to apply restrictions on further logging and road building to all of Alaska's roadless lands. Dombeck explained the key decision criteria to me like this: "Alaska's roadless lands are the best we have. We need to protect the best."

His logic had a simple, profound elegance that perhaps I failed to appreciate at the time. There exists much nuance between the poles of protection and wise use of lands. The choice is almost never clear, and claims are often mutually exclusive. I found the arguments of both sides compelling. Alaskans do rely heavily on the Tongass for the basics of life. Ecologically, the Tongass is of global significance. That's what makes this a "wicked" dilemma and why I found it difficult to see this clearly at a crucial moment.

The rush to complete the regulation continued into late fall. Election Day dawned. Like the rest of the country, we were riveted to the news announcements of election returns. It's Bush...It's Gore...It's Bush again...Chads are hanging in Florida, then the drama moves to the Supreme Court, which decides in a five-to-four vote to confirm George W. Bush's victory over Al Gore.

In early December, we published the final Environmental Impact Statement and the proposed Roadless Area Conservation Rule, with a thirty-day waiting period before it became effective. George Frampton, chairman of the Council on Environmental Quality (CEQ) and the president's leading spokesman on environmental matters, chaired a briefing for all federal agencies interested in the roadless proposal.

The Department of Energy was one very interested agency. The Roadless Rule proposed to ban any more road building in connection with oil and gas leases, beyond roads that existed or those needed for current leases only. North Dakota's Little Missouri National Grassland held a potentially huge but relatively undeveloped gas field for which restrictions on road building would make development, if not impossible, much more costly and difficult.

As the briefing began, I thought, "Now we have the Day of the Suits." A contingent of Department of Energy officials made an impressive entrance. The one woman among them wore a bright red dress, exuding power.

The leasing issue quickly arose. Frampton laid out the premise of the regulation on leasing. The folks from the Department of Energy weighed in using strong, unequivocal terms: this had to be changed. Frampton countered that the prohibition on new roads was a well-considered element of the rule. Acknowledging an explicit trade-off, Frampton said it was needed to protect roadless lands from the probability of more "dry holes" that would achieve little except to ruin more precious roadless lands.

The DOE contingent ratcheted up the argument, saying it would not accept any outcome other than making a last-minute change to the regulation.

It seemed clear to me that this conflict would have to come before White House Chief of Staff John Podesta and maybe President Clinton himself. And sure enough, in the next few days the regulation was rewritten to exempt any renewed lease from road-building restrictions. DOE won the argument, while CEQ gained some solace by retaining restrictions for all future leases. The lands in question later became a centerpiece of a resurgent oil and gas industry in North Dakota.

The final rule was published in the Federal Register on January 6, 2001. Two weeks before George W. Bush assumed the presidency, William Jefferson Clinton, clad in a black overcoat, proudly made a presidential announcement at the National Arboretum, with gently falling snow dappling his shoulders. It seemed a perfect ending to a strenuous but deeply satisfying effort.

I reflected on all that happened in the past eighteen months. I considered it a privilege to be given the opportunity to fashion regulatory protection for roadless areas. Nearing the end of my career, I had helped fashion a sweeping, bold policy for the environment. The same agency that had been logging and building a vast network of roads into natural landscapes wheeled about-face to strive for durable protection. Protection from what? From the Forest Service itself.

Ecological Sustainability

Meanwhile, our concurrent effort to complete the National Forest Management Act (NFMA) planning regulation was moving ahead on schedule. Though not as politically charged as the roadless issue, the planning rule had many critics, among them many Forest Service employees. We believed that, like the roadless rule, the planning rule risked being scuttled by the Bush administration unless we could complete the effort during Clinton's term.

The previous planning rule dated to 1982 and was heavily weighted toward timber production. It badly needed an update to more accurately represent changes in public values. We also hoped to include improvements in Forest Service planning technology and procedures, such as powerful geographic information systems (GIS) that were largely unknown in 1982.

A committee of scientists (COS), headed by Norm Johnson, conducted an exhaustive review of agency practices and produced a report titled *Sustaining the People's Land*.[16] I found it fascinating that, in contrast to an earlier COS that had developed the conceptual thinking for the original NFMA planning process, Johnson's new team came up with vastly different recommendations for the very same NFMA.

They recommended that sustainability, with its ecological, social, and economic elements, be the "guiding star" for Forest Service planning. Of these elements, the committee singled out ecological sustainability as the foundation from which social and economic sustainability flow. Think of a three-strand rope. The COS saw social and economic sustainability as related to but distinct from ecological sustainability. The COS believed that the various elements of sustainability could be unraveled into three component strands, but the ecological strand was the highest priority for the Forest Service because the agency's fundamental responsibility was to manage land and resources.

The original spirit of the NFMA was intended to curb logging excesses on our national forests. The new COS report conceptualized national forest reforms, long overdue, that would foster greater attention on sustaining entire ecosystems and revise the timber-centric forest plans that had been written throughout the past two decades.

As the planning rule took shape, three prominent issues emerged. First, how would the Forest Service implement the concept of ecological sustainability? Second, could the Forest Service successfully disentangle itself from the legal requirement of the 1982 regulation to "ensure the viability of vertebrate wildlife populations"? Third, could the timber production emphasis, which drove almost all the Forest Service's on-the-ground land management, be reoriented to complement and achieve ecological sustainability principles?

The concept of ecological sustainability sparked passionate discussions among the Forest Service's larger scientific community. The Forest Service has the largest forest research organization in the world, and, even though Johnson's committee included mostly esteemed biological scientists, a few prominent Forest Service scientists began to rally in opposition. They maintained that there was no such thing as ecological sustainability apart from social and economic sustainability. In short, they believed that the three elements were inextricably linked and could not be unraveled. If separated, each element had no individual identity or integrity.

Early drafts of the planning regulation endeavored to weave the COS's concept of sustainability into the basic foundation and philosophy of forest planning. Keeping the concept of ecological sustainability at the forefront made us change the way we asked and answered important questions like "Why plan?" and "What are we aiming for?"

One of my favorite adages is, "There is no right way to do the wrong thing." I understood that some of our Forest Service research scientists clearly felt our approach was the wrong thing.

Even so, Jim Lyons, the under secretary of agriculture, strongly supported the COS's position, as did Chief Dombeck and I. Pete Roussopoulos, Southern Research Station director, and Ariel Lugo, director of the International Institute of Tropical Forestry, led the opposition, cheered on vigorously by their boss, Robert Lewis, deputy chief for research.

What I found most bizarre throughout several heated discussions was the lack of engagement by the regional foresters. These nine leaders would each oversee completion of numerous forest plans based on the planning regulation. They clearly had the most at stake. For the better part of the discussion, they sat stone-faced, apparently bored by the debate. Given the gravity of the issue, I was a bit irritated at their diffidence.

Forest Service researchers produced a lengthy position paper methodically laying out their opposition and explaining why they viewed the COS's concept as invalid. More debate yielded no meaningful compromise. Jim Lyons and Mike Dombeck finally said that, with all due respect to agency researchers, they had heard enough. They signaled their intent to move on with the new planning regulation, reflecting their determination that ecological sustainability was, in fact, unique and valid and would be the Forest Service's guiding star.

Determined dissenters promptly authored yet another formal policy paper refuting the COS's concept. I recalled a bumper sticker that read, "The part of 'NO' I don't understand is where I don't get what I want."

Jim Lyons was thoroughly fed up with the bickering and insubordination. I sympathized with him. He decided to put the issue to bed and quickly convened a meeting in Oregon with several noteworthy scientists, including some Forest Service field researchers. He posed a simple question: Can ecological sustainability be considered separate and distinct from its related social and economic sustainability elements?

Their answer was yes, issued in a brief white paper back to Lyons. Done, in Lyons's mind.

Next issue: viability—the crucial lever that had turned the spotted-owl controversy in favor of environmentalists. The NFMA includes a goal that forest plans "provide for diversity of plant and animal communities." The 1982 planning regulations specified that the Forest Service had to maintain "viable populations" of certain species. Environmentalists had repeatedly held the agency's feet to the fire on this in court, but in truth it represented an intolerable burden, especially in light of the timber focus in the 1982 planning rules. The viability provision made forest managers legally liable for a difficult balancing act.

Historically, the Forest Service confined its responsibility to that of managing habitat. The responsibility to manage wildlife populations rested with states or with other agencies, such as the Fish and Wildlife Service. Estimating populations is expensive and difficult. It is also highly uncertain: many things can happen to wildlife that are unrelated to forest management—migratory birds can die in South America, elk can be poached, trout can die of whirling disease.

We intended to craft the 2000 planning regulation to comply with the diversity requirement of the NFMA, but eliminate the legal liability inherent in the viability requirement. Accordingly, our proposed new planning rule made managers responsible only for wildlife diversity. This triggered fresh fury from the environmentalists who did not want to lose such an important legal advantage.

Even though the environmental community generally supported the Clinton administration's policies, they remained suspicious of the Forest Service and were especially wary of leaving any door ajar that could be exploited by a future unfriendly Republican administration. In this case, they insisted that the Forest Service stand by the viability requirement. When we issued our draft planning regulation without it, the fight was on.

Jim Lyons supported dropping the viability provision as long as we diligently pursued a rigorous effort to ensure diversity in the new regulation. Our wildlife experts, notably Chris Iverson, worked tirelessly to thread the needle, but Lyons still caught hell from environmentalists.

The third issue—ecological sustainability as opposed to timber production—struck me as the most important in terms of scope and future implications.

The NFMA, when it was enacted in 1976, was a significant advance in environmental policy because it argued that the singular pursuit of wood products on national forests was antithetical to the reason national forests were created in the public interest. In practice, however, timber harvest became the dominant objective in the first generation of the forest plans that the NFMA required.

My responsibility for management policy on all national forests carried a duty to ensure that the new NFMA planning regulation complied with both the NFMA itself and the committee of scientists' findings and recommendations. In the past fifty years, the Forest Service had managed more land through timber sale contracting and had spent more money on it than on any other activity. It was essential that we now develop a framework for cutting trees primarily for reasons other than timber production.

I mentioned to a few colleagues how important I thought this issue was. Sally Collins, my associate deputy chief, seemed the only one who agreed or cared much. Undaunted, I authored language in the regulation that was intended to move away from the narrow focus on commercial timber production, with the goal of transitioning the Forest Service to a broader goal of sustainable timber management primarily for the purpose of restoring forests to more natural conditions.

The final planning regulation was published in the Federal Register on Monday, November 6, 2000—one day shy of Election Day. Completing both the Planning Rule governing all 155 national forests and grasslands and the Roadless Area Conservation Rule for fifty-eight million acres constituted two tremendous accomplishments.

The twin victories enjoyed a brief life span. Americans elected George W. Bush, inaugurated into office on January 20, 2001. This changed everything.

Chapter 10: Over and Out

WASHINGTON, DC, PART IV, 1999–2002

After the Supreme Court drama of the Bush-Gore election settled down, another drama invaded the suite of Forest Service executive offices. Rampant speculation had Chief Dombeck being fired or quitting soon after Bush took office. The chief did nothing to dispel the speculation. The rumor pipeline ran full with theories on who would succeed him to become the fifteenth chief of the Forest Service.

Besides Chief Dombeck, the axe hung over others, including me.

Hilda Diaz-Soltero, an associate chief and my boss, would be on the hit list with me because of our prominent role in developing the Roadless Area Conservation Rule. Diaz-Soltero had an additional problem: having come from the Fish and Wildlife Service, she was an outsider to whom traditionalists within the Forest Service owed nothing. Although no action would come immediately, leadership changes in a few months' time would reveal the ever-increasing politicization of the agency.

Phil Janik, the other associate chief, also felt insecure, owing largely to his earlier actions as Alaska regional forester. The Alaska congressional delegation, empowered by a friendly Republican administration, would expect Janik's departure.

Lyle Laverty, Rocky Mountain regional forester in Denver, quickly emerged as the odds-on favorite to replace Dombeck. He certainly lobbied aggressively for the job. His appointment calendar for December 2000 through February 2001 would make an interesting read. Rumor had it that Laverty spent a lot of time in DC visiting influential senators and representatives as well as new Bush appointees to assert the wisdom of his succeeding Dombeck. Laverty, a traditionalist, had a strong background in recreation and, more recently, fire issues. His selection as chief would signal that the Forest Service would revert to more conservative policies than those of the Dombeck era.

Other regional foresters appeared to be lying low. But by late February and early March the name of Dale Bosworth, Northern Region forester in Missoula, Montana, began to surface regularly. Bosworth's father had a long Forest Service career, and his son also joined the agency. Three generations! My National Forest System directors in DC regarded Bosworth highly, as did his peer regional foresters and, I'd guess, almost anyone who knew him. Bosworth was a classy professional and a well-respected man.

Bosworth often remained quiet, or at least reserved, in tough policy discussions at our national leadership meetings. I had difficulty getting a firm read on how he felt about things. When he did speak, his positions struck me as nuanced and lukewarm—neither strongly for nor against much. Did his motives put a priority on playing it safe? Not showing his true heart?

After Jim Lyons departed as under secretary of agriculture, Dave Tenny, who previously served on the House Agriculture Committee staff, stepped in for several months on an acting basis while the eventual Bush nominee for under secretary, Mark Rey, sat and awaited a lengthy Senate confirmation process. Rey was rumored to prefer a prominent post at the Office of Management and Budget, which he did not get. Rey, staff director of the Senate Energy and Natural Resources Committee, had deep ties with the timber industry and a strong working knowledge of the Forest Service.

On inauguration day, the Bush administration immediately suspended all recently completed federal regulations. For the Forest Service this held enormous significance—both the roadless-area and forest-planning regulations were caught in the net and would not take effect. I was demoralized, though not surprised. I knew that a Republican administration would swing right, and Bush wasted no time sending a strong signal. This was hardball.

Mike Dombeck had occasional meetings with Ann Veneman, the new secretary of agriculture, who was from California. He reported his interactions as respectful, cordial. But a strong undercurrent of tension attended their relationship. I imagined each looking at the other, thinking, "You are not part of my future."

In order to keep the wheels in motion, Dave Tenny immediately brought a few Forest Service staff into the under secretary's office to deal with the press of business that rushed in. Among these were Chris Risbrudt and Bill Sexton, director and assistant director, respectively, of my Ecosystem Management section. They had been charged with implementing the new planning regulation. Might stripping my office of key staff be an omen? Yes.

In addition, as expected, numerous parties filed lawsuits aimed at both the roadless-area and forest-planning regulations. Idaho filed an immediate suit against the roadless rule. Although we knew of this action, we heard nothing from the Department of Justice. Normally, people like Dombeck, Diaz-Soltero, and me, who all played prominent roles in the creation of the regulation, would be asked by Department of Justice attorneys to provide a deposition in preparation of the government's case.

Imagine our collective surprise when the first hearing arrived and the Department of Justice offered no defense. I'm not talking about a weak effort or sloppy work. They offered nothing. They never discussed the lawsuit with Dombeck, neither how it intended to represent him, nor his legal liabilities as a named defendant. The Department of Justice offered no argument on behalf of the government.

Dombeck viewed this as the tipping point he'd been waiting for, a provocation that would serve as a just cause to resign. On March 27, 2001, Dombeck wrote a letter of resignation to Secretary Veneman. The letter included a lofty defense of his record and the accomplishments of the Forest Service and alluded to the risks and consequences of an about-face on important land-ethic principles such as roadless-area protection.

Dombeck felt hung out to dry by the Department of Justice and did the honorable thing by resigning. This suited the Bush team—they wouldn't have to fire him. Chris Wood, Dombeck's close confidant, soon followed him out the door.

Meanwhile, the Bush administration, in a bit of a rush now, cranked away at finding a successor to Dombeck. News accounts of his resignation named Bosworth and Laverty as possible successors, indicating that rumors about the two front-runners were accurate.

Then, within a few days of Dombeck's departure, Secretary Veneman named Bosworth as the next chief. Rumors indicated that Bosworth had not sought the job, but that he would be honored to serve if asked. Laverty had made it clear that he wanted the job badly—too badly, in fact. Laverty had overplayed his hand. His lobbying for himself established him as the early favorite, but the more humble and patient Bosworth had prevailed. The outcome seemed to validate the Forest Service's old-guard culture; the tortoise beat the hare again.

Hail to the Chief

Symbolically, perhaps, my third-floor corner office sat directly below the chief's, and issues could be shoveled down the chute to land on my desk. I enjoyed a close working relationship with both Dombeck and Wood. One day, while all these machinations were going on, Wood popped into my office just to remark that he was glad I was the deputy chief. I had to ask, "What prompted that?"

He said, "When you and I discuss things, after I leave, you do what you said you'd do. It wasn't always that way." This simple gesture gladdened my heart. It made me sad to think it could be any other way.

I considered the opportunity before me to work with and for Dale Bosworth. I asked for a meeting, just the two of us.

We'd known each other casually for many years. I remembered when we'd gone charter fishing off Ocean City, Maryland, in about 1990. Dave Rittenhouse caught a nice seventy-pound yellowfin tuna on that trip, and I had my first taste of fresh tuna hot off the grill—really fine dining. Bosworth and I both had attended numerous meetings of the agency's leadership team. Our relationship was not close, but it was professional and respectful. I thought we got along well.

In fact, Bosworth had received some help from me a few months before he became chief. The USDA's Office of the Inspector General (OIG) had conducted an investigation involving land transfers and sales in Nevada and concluded that serious improprieties had occurred, resulting in the loss of ten to fifteen million dollars. Bosworth had been Intermountain regional forester at the time of the event. The matter was under his jurisdiction and USDA sought to hold Bosworth culpable.

I got involved along with Hilda Diaz-Soltero because the issue was a national forest matter that involved our subordinates. We defended Bosworth vigorously and successfully. I disagreed with OIG's assessment and conclusions. Based on my reading of the USDA's investigation, Bosworth had played a peripheral, nonmaterial role. Had OIG's view prevailed, the incident could have damaged Bosworth's reputation and his future with the Forest Service.

I defended Bosworth because I thought his actions, more a lack of effective oversight, did not warrant severe punishment. I wondered whether he might defend me when and if the push came to remove me. I wasn't sure he even knew of my role in backing him, however. As it turned out, others in the Nevada land affair were deemed more culpable; several were disciplined severely, and some retired.

My face-to-face visit with Bosworth turned out to be a cool, matter-of-fact, quite blunt discussion. I shared my view that Dombeck had accomplished things that were important to him because he surrounded himself with people he could trust. I said I had served Dombeck faithfully and loyally. I always strived to tell Dombeck what I thought. I told Bosworth that I intended to do the same for him. I knew he had the prerogative to bring in a new deputy chief, and I wouldn't kick if he did. I said we both knew that the circumstances of my becoming deputy chief were unusual at best, and that they had offended some people. Bosworth and I both knew that he was much more the traditionalist than Dombeck had been.

In fact, Bob Jacobs, Eastern regional forester in Milwaukee, Wisconsin, and one of our fellow Ocean City fishermen, had confided to me Bosworth's "outrage" when he heard I got the job. Jacobs quipped that Dombeck had first offered Bosworth the deputy chief job after Gray Reynolds left, but Bosworth didn't want it. Jacobs had asked Bosworth to give me a chance; at least I wanted the job.

Bosworth simply said that no immediate changes were forthcoming. He said he just needed the deputy chief's office to do our job. Fair enough.

Or so I thought. I soon noticed that important as well as routine national forest issues and policies were bypassing me and my staff entirely. I suspected the culprits worked in Dave Tenny's office. I brought the issue to Bosworth's attention. I said I had no quarrel with the prerogatives of USDA (Lyons had also been an activist, hands-on under secretary), but we were not even being informed, let alone involved, on national forest policy changes.

Case in point: our field organization needed clarification on how to handle a few things like Christmas tree cutting and firewood gathering in roadless areas. A memorandum had been prepared by Chris Risbrudt, now in Tenny's office, to distribute to all the regional foresters, who were in town for a national leadership meeting. I'd had no input. I mentioned to Bosworth that I'd seen the policy memo and thought it was going to create confusion and misunderstanding. He acknowledged that Tenny's office should be working with us, not excluding us. This was inappropriate and not in line with his wishes. He directed Risbrudt to confer with me before distributing the directive to the field.

Risbrudt gave me the memorandum, and we agreed to meet for a discussion the next morning over breakfast. I edited the memo on the subway ride home that night. In the morning, I discussed my suggested changes. No substantive discussion followed. Risbrudt listened, picked up my edited memo and left.

Risbrudt distributed the memo to our leadership team right after lunch. I quickly scanned the letter—not one change. I got the message.

Within thirty seconds, a regional forester stood up and pointed out something in the memo that didn't make sense, then another, and another. Each point had been explicitly addressed by the edits I'd provided Risbrudt. The memo was a mess. Bosworth, embarrassed, quickly asked to pull back the memo and pledged to get it fixed. Risbrudt retrieved all the copies. The rereleased memo came out shortly after with most of my fixes. I told the chief how this went down. I thought we could do better. This certainly didn't make the chief look good. He deserved better.

Things got worse. I clearly had no legitimacy—nothing came through me, nothing came from me. I was marginalized, irrelevant. I found it tempting to take the disrespect personally, but I knew that, professionally speaking, this was the harvest of prior planting in Dombeck's acreage. Though the treatment was not unexpected, the rapidity with which it happened surprised me. I could understand the sentiment behind it— Dombeck had taken the Forest Service in what was now judged to be the wrong direction, and I was too eager to help him. Yet it still stung.

My predicament had implications not just for me but for the entire agency. I was the deputy chief, but I was of little use to the chief. Not much was asked of me in managing national forest affairs. The Forest Service needed an effective agent for national forest policy. I also believed that Chief Bosworth needed a deputy he believed in, a person he could trust without reservation. Obviously, I was not this person. I knew how situations like this ended.

It was time to start weighing my options.

My situation was crystallized in an episode involving congressional testimony for the agency, which I frequently provided. I met with our legislative affairs staff to prepare to testify before a House subcommittee on a relatively innocuous bill. The next morning, as I got ready to go to the hearing, Tim DeCoster from Legislative Affairs informed me that another person would be going in my place.

That surprised me. No explanation was given. The few hours of preparation seemed a waste.

A couple of weeks later the same thing happened, but this time notice came the prior afternoon. At least I didn't come to work expecting to go to Capitol Hill. Once again, there was no explanation.

Then it happened a third time. It was time to talk with the chief.

Bosworth seemed surprised, saying this had not been his decision. Hearing this, I told him the pattern pointed to an unmistakable conclusion: that the USDA had intervened to ensure Jim Furnish would no longer represent the Forest Service before Congress. I said I could live with that, but preferred not to speculate. Could he confirm this with the USDA? If true, could he also ask why and let me know the reason?

I had strong suspicions that Dave Tenny made these decisions. Tenny had served as staff for a House committee prior to his USDA appointment. I'm sure he saw me testify numerous times. I consistently supported Clinton administration policies, but in no case did I take a position I didn't believe in. Yet I could understand that Tenny might take me to be a loose cannon.

At this point, I'd had two meetings with Tenny. When Secretary Veneman took office, it seemed everybody in her home state of California with a gripe about the Forest Service had deluged her office asking for relief. Tenny asked me to come and see him about a conflict involving television transmission towers in the foothills outside Los Angeles. Univision, the Latino media company, claimed that the Forest Service had given ABC permission to install a new tower that Univision believed would interfere with its transmission capability. Letters were flying between lawyers. Tenny wanted me to straighten this out "yesterday."

I said I'd be happy to work on it, but there would be no instant fix. I would have to work through the conflict with the staff on the Angeles forest, and the Federal Communications Commission would necessarily need to be involved because this involved a technical matter. We'd have to get the lawyers calmed down and talking to one another to resolve the issue. This would take a few days—I estimated about a week. He seemed irritated.

I bristled a bit and said that if he wanted me to resolve the issue, he had my honest assessment of what it would take. If he wanted it solved in a day, he could do it himself. He gave me the folder. It did take a week, but the issue resolved to everyone's satisfaction.

On another occasion, Tenny wanted to see Hilda Diaz-Soltero and me for a talk. When the three of us met, Tenny laid out his expectations, asking if he and the Bush administration could expect our support for their positions on Forest Service matters. I said this seemed reasonable as long as I could speak freely to him on important policy matters. Once he made a decision, he had my support, unless I could not live with his position. In that case, if the issue carried enough weight, I could resign.

A couple of days after my meeting with the chief about the mystery of canceled testimony, Bosworth asked to see me again. He said he had met with Dave Tenny, and the decision was firm: I would no longer testify. I thanked him, saying it would save everybody wasted time and energy and give clarity to legislative affairs in lining up witnesses.

Then I said, "Why doesn't Tenny want me to testify?" Bosworth said, "He doesn't trust you." I told him that Dombeck had appreciated my loyalty. I said I would continue to be loyal if given the opportunity. Try me. He said the decision was final.

"I can understand that Tenny doesn't trust me," I said. "What about you?"

No answer. I told him that, when he accepted Tenny's view and declined to defend me as the agency's deputy chief, I took this as professional disrespect and a signal of his own distrust. I left, keenly disappointed.

When I had departed Corvallis for Washington back in 1999, I recognized that my tenure as deputy chief probably hinged on the 2000 election. Now, I found, none of these discouragements really surprised me. I was witnessing my own demise with an almost morbid curiosity. So...this is how it happens?

The next few weeks felt like a delicate dance with the wrong partner. When Mike Dombeck became chief, he made no secret that certain executives were no longer needed, and he swiftly made changes to surround himself with people he trusted to accomplish his agenda. It seemed clear to me that the same dynamic was in play now, but nothing was actually said. But no one need say anything; I no longer felt wanted or needed.

I had too much respect for the agency to occupy a position but not do the job. Management of our national forests needed a leader, not a toothless figurehead. The awkwardness grew for Bosworth and me. I figured the day would arrive when he would summon me in and tell me my career was over.

Enter Sally Collins, selected to become the new, single associate chief after Bosworth had unified the duties formerly split between Phil Janik and Hilda Diaz-Soltero. Collins came to talk to me in my office. She delicately introduced the topic of a new deputy. I said I understood perfectly. Let's deal with reality.

We agreed that the current situation was untenable. Collins encouraged me, saying I had so many good ideas and such good energy, would I consider moving to a different position? Both Janik and Diaz-Soltero had accepted demotions. Diaz-Soltero had been reassigned to direct the Pacific Southwest Research Station in California, while Janik chose to remain in DC, buried deep and of no influence any longer.

I thanked Collins for the kind words. I thought I did have good ideas, and I still cared passionately about the Forest Service. I still had much to offer if Bosworth would use me, and she agreed—but not as his deputy chief.

At this point the conversation got sticky, intense. I said I didn't come to Washington to hang around and draw a paycheck. We both knew that I was Dombeck's choice as deputy chief. If Bosworth no longer trusted me, no longer needed me, I didn't see any benefit to the agency or myself in taking a lesser position to simply pad my retirement annuity. That seemed wrong and offensive on many levels. It symbolized all that's wrong with bureaucracy. If the chief didn't want me, just tell me so. Let me go.

I think Collins found me difficult. I thought I was being easy. And my solution would be a whole lot cheaper than hers.

Bob Jacobs visited next. Bosworth had brought Jacobs in as chief of staff from his regional forester position in Milwaukee, Wisconsin. I'd worked for Jacobs on my first tour in DC; he'd been an important mentor. We had a good relationship and spoke the truth to one another.

But he was now quite coy. He gave me the impression that a change was expected, but the next move was mine. If I voluntarily stepped aside, they would find another job for me, and I could play out the string until I chose to retire. I sensed that Bosworth did not intend to oust me. I still didn't understand why, but I guessed that he preferred to move me down in rank to a position where I could do no harm.

What if I chose to play hardball and stay put? Too many assumptions floated about. I tried to be direct and unequivocal. It seemed to me I got mushy evasions in return. I think they calculated that a person of integrity would "do the right thing" and vacate the deputy chief's position, opening the way for the chief to fill the job with a person more to his liking. They were willing to take that risk with me, wagering that I'd voluntarily step aside soon. I told Jacobs I'd think about the situation and let him know.

Thirtymile to Go

When the phone rang, I glanced at the midnight hour on the bedside clock. Good news didn't usually come in the middle of the night.

"Hi, Jim, this is Carolyn Holbrook, from Safety." My heart sank immediately. "We had four fatalities on a fire in Washington. I'm asking if you'll head up the fatality investigation." It was July 10, 2001.

The Forest Service itself seemed engulfed by fire. The Storm King Mountain fire near Glenwood Springs, Colorado, had killed fourteen in 1994, and several other fatalities in ensuing years, some involving aircraft, hit the agency hard. The cost of fighting fires also spiraled out of control. These four new fatalities were a serious punch in the gut.

A few months earlier, we senior executives got reamed out by Chief Dombeck for dodging the responsibility to lead such investigations. We had an executive roster and names were progressively rotated to the top for two weeks at a time, then back to the bottom of the list. In a drowsy fog, I recalled that my slot was July 1 to 15, and this was why Holbrook was calling me. The last time they needed an investigation leader they went through seventeen names and seventeen refusals until someone said yes. We were told that this was absolutely unacceptable. Investigations were rarely needed, but we were expected do our duty when our time came.

I immediately said yes.

I met Chief Bosworth first thing in the morning at our long-scheduled National Leadership Conference in DC. I briefed him on the fatalities and my departure. Then I excused myself and headed for Reagan

Airport and boarded a plane to Winthrop, Washington, where two men and two women, all young, had perished in the Thirtymile Fire on the Okanogan National Forest.

The investigation was an all-consuming assignment from July 10 until late September. It was tough, sobering, difficult work. I've come to view the investigation as among the most important things I did for the Forest Service. It was also the last thing.

These four firefighters died tragically, unnecessarily. The investigation team observed a thin margin between life and death—twelve other people at the scene survived, including two civilians. The death toll could easily have been much higher and more horrific.

As our investigation team convened in Winthrop on July 11, a tense mood permeated the conference room. Expectant media thronged the lawn outside. The Forest Service had often been accused of white-washing fatalities, improperly absolving people of wrongdoing. I remembered that, after the Storm King fire, one of the investigation team had refused to sign the report for this very reason.

I told our team at the outset that we would conduct an honest, thorough, and unflinching investigation. I vowed that everyone on the team would be able to sign the report with no reservations if we did our job well.

When the fire descended on those sixteen people huddled in awe and fear along the Chewuch River, most survived, in spite of all the mistakes that were made. Personal choice spelled the difference between life and death. When the wall of fire slammed into the group, most of them deployed their fire shelters on a two-lane gravel road. The two civilians crowded into Rebecca Welch's one-person shelter, and all three survived.

But six firefighters deployed their shelters in various locations above the road, on a scree slope of large boulders and amid a scattering of trees. Thom Taylor, isolated a short distance uphill from the others, survived the initial firestorm under his shelter and then made a break for the river fifty yards downhill. Taylor dove into the water with his shelter trailing like a silver shawl. He lived.

Jason Emhoff briefly remained under his shelter. But, after trying with bare hands to extinguish woody debris that was on fire under him, he feared he would perish and bolted downhill to the road. Jason sought protection inside the fifteen-passenger van parked on the road and survived, albeit with severe burns on his hands.

Tom Craven, Karen FitzPatrick, Jessica Johnson, and Devin Weaver all perished in their fire shelters, asphyxiated by super-heated air. Our investigation report concluded that twigs and branches lodged among the boulders had already ignited by the time they got inside their shelters. This "fire within," coupled with the super-hot fire outside that surged up through the large boulders, spelled doom.

For reasons that remain murky, six firefighters did not join their crewmates on the road but chose to remain on the slope above the road. That location became lethal for four of the six when the firestorm arrived. An unspeakable tragedy ensued; just a few yards away, everyone on the road survived.

Our team concluded that the tragedy occurred because of an agency fire culture that fails to lead people to judiciously balance risk and prudence. Earlier in the day, the fourteen firefighters became trapped as the fire burned across the road, cutting off the only way out of the canyon so they could not retreat to safety. Long before this critical juncture, Thirtymile Fire had already escaped containment and blown up far beyond the capacity of this small crew to make any measurable difference in the outcome. The harsh reality of hindsight indicated that the crew needed to accept the fact that the fire was lost, retreat, and wait for a large-fire Type I team to assume responsibility.

Our investigation report was signed by all team members. But our conclusions were controversial, in part because we assigned a portion of the responsibility for the deaths of the four to their choice not to join those on the road, though asked to do so. Most of the responsibility, however, we assigned to poor leadership at the scene: notably, the fire captain who had failed to make sure everyone in the group was gathered in a safe place to face the approaching fire, prepared and ready for the worst.

Chief Bosworth later revised our findings to absolve the four victims of all responsibility—a conclusion our team had rejected in the course of our investigation. In his book *The Thirtymile Fire: A Chronicle of Bravery and Betrayal*,[17] John McLean quotes the chief as saying that he'd learned one lesson from other fire fatalities: "Never blame the victims." I can understand his compassion, but modifying our report to accommodate this sentimental view was not supported by the facts.

The Forest Service developed an outstanding fire safety module based on our investigation of actions in the face of danger. I hope future firefighters will survive because they exercise greater prudence when balancing risk and taking action. No one should have died. As in almost all fire fatalities, poor choices and failures in judgment along with violations of standard fire safety procedures directly contributed to the deaths.

Walk a New Walk

As the fire investigation wound down, I continued to weigh my options about the deputy chief situation. I had actually prepared myself for this circumstance before I left Corvallis, and now it was real. I knew that I had become a deputy chief because of my pro-environment record, and now I was losing the job for the very same reason. I wanted to stay, but not under the prevailing conditions, which I viewed as unlikely to change. Chief Dombeck had given me the opportunity of a lifetime, and I think we achieved great things.

I had spent my professional life investing in the Forest Service. It now seemed clear to me that if I cared about the organization more than myself, it was time to leave. I had accomplished more than I could have ever dreamed, and I had few regrets. I had no desire to start accumulating resentment.

In late August, I requested a meeting with Bob Jacobs. I came directly to the point, telling him that the fire investigation was nearly complete and I thought departing soon best served the agency; Bosworth needed a new deputy chief. Jacobs listened, offering nothing in return.

"About the earliest I feel I can leave would be October," I told him, mindful that I would turn fifty-six on October 3. Jacobs then simply said, "Good. I'll let the chief know. Let's plan on that."

I was curious whom they had in mind to replace me. "Can't say," he said.

Joe Walsh, in public affairs, asked to see me a couple of weeks later. He showed me a press release announcing my retirement and wanted to know if I approved the wording. It was pretty straightforward boil-erplate, stating that I was "retiring from the Forest Service" after a long career. I took my pen and struck "retiring from" and wrote in "leaving."

Walsh smiled and said, "That's stronger!"

I said, "Yes, and also truer." Walsh said he'd have to run it by the chief. If Bosworth agreed to the change, they'd release it to the media. If not, Walsh said, "You and the chief will have to talk." I said it was a small but important matter to me, and I welcomed the chance to talk with the chief if need be.

About a half hour later my phone rang. The chief wanted to see me. That didn't take long! I rehearsed my thoughts as I made my way up the stairs to his office. He met me outside his office door with papers in hand. No invitation into his office for a sit-down.

He pulled a letter from a folder. The brief letter assigned me to a ginned-up Thirtymile fire coordinator position effective October 10, allowing me occasional work as necessary to attend to fire issues while I utilized available vacation time until my official departure date in early January 2002. If I agreed, I was to sign at the bottom. I signed the letter. Bosworth made no mention of the press release.

The press release went out with my change intact, simultaneously announcing the selection of Tom Thompson, Rocky Mountain deputy regional forester, as my replacement. Both he and I had earlier been Siuslaw National Forest supervisor.

In a better world, I like to think Thompson and I would have been close professionally. In truth, he represented an ideology that con-trasted with mine in about every way possible. Thompson's selection as my successor implied that things were back to normal.

In December I did a two-hour-long on-camera investigative inter-view with Keith Morrison of ABC News about the Thirtymile Fire fatalities. Chief Bosworth granted Morrison a half-hour interview. ABC never aired the show.

After my speech at a good-bye reception in the USDA atrium, I took off my dress shoes, dropped them in the trash and walked to my car in stocking feet with Judy for the trip home. The time had come to walk a new walk.

I retained an interest in the legacy we'd created, primarily roadless protection and the planning regulation. Both endured an adventurous and bumpy ride, long in reaching a conclusion.

The roadless regulation was litigated on many fronts for many years, remaining undecided until February 2012. A three-judge panel of the Tenth Circuit Court of Appeals in Colorado, including two George W. Bush appointees, ruled unanimously in support of the 2001 Roadless Area Conservation Rule, overturning a ruling by Wyoming District Court Judge Clarence Brimmer. The Supreme Court refused to hear the case on appeal and let stand the Tenth Circuit's ruling.

Reading their opinion was sweet vindication. After all these years, the judges strongly supported our work on all counts. Delayed gratification washed over me as I thought back on our work for the cause of roadless-area protection. We did the right thing. We did it the right way.

The Bush administration suspended the planning rule in January 2001. The Forest Service completed the burial by dropping the planning rule entirely, saying that the agency would create a new version. The agency released a new planning regulation in 2005, and litigation followed immediately. Beaten soundly in court, the Forest Service tried again, releasing yet another version in 2008. Litigation again followed.

When President Obama took office, Secretary of Agriculture Vilsack announced that the Forest Service would drop the 2008 planning rule rather than pursue further litigation. A sweeping, comprehensive effort culminated in the release of a new 2012 planning regulation, which has an uncertain legal fate. I judge the 2012 planning rule as much more similar to the 2001 rule than was the 2008 version—actually, better— and the 2012 version enjoys better support from all affected interests and is well regarded by agency planners who will implement the rule. I hope it works. I think it will.

Chapter 11: A Green Manifesto and Our Future

When I arrived in Tiller, Oregon, in 1965 for my first Forest Service job, I met my boss for the summer, Owen Downhill, Jr., who lived in a small house on the Forest Service compound with his wife and young children. Downhill very graciously cared for us, his young, inexperienced crew, while we learned the ropes. I had no car, so I walked around town to the two big attractions, Ruth Mount's 42 Cafe and Porter's grocery store/post office.

Downhill knew I loved to fish, so he took me to Squaw Creek, one of his favorite spots. Early on a Saturday, we followed the South Umpqua upriver to its confluence with Jackson Creek (where I once saw an otter lazily schooling a bunch of suckers), then, passing near the world's largest sugar pine, took a logging road a few miles farther southeast, heading higher into the mountains. We parked on the shoulder of the road in a fresh clear-cut harvest. Heading straight downhill, crazy steep, we followed a crude firebreak built by a dozer. Charred logs lay scattered about from when the Forest Service burned the clear-cut to get rid of logging debris so the site could be planted more easily with fir seedlings. It was sunny and hot.

We quickly reached Squaw Creek, no more than ten feet wide, tumbling down through giant trees, a never-ending series of crystal-clear pools amid the boulder-strewn channel. Native cutthroat trout awaited us. Lost in reverie, we fished for hours, leap-frogging our way slowly in a gentlemanly fashion, making small talk occasionally. We caught and released seven- to eleven-inch trout too numerous to count, each one beautiful and, in a humans-as-novelty way, quick to grab even the most sloppily presented fly. What a great day. We sweated buckets on the trudge uphill to the car, a small price to pay.

Thirty years later, I met Jeff Dose, a fisheries biologist for the Umpqua National Forest, at a Forest Service meeting. Dose got crosswise with his boss because he'd gone rogue, speaking openly and publicly of his view that timber harvesting ruined stream health and fisheries. This earned him much scorn from agency regulars and kudos from the environmental community, who knew there was abundant evidence of environmental trouble if they could only find it. Dose's outspokenness fueled suspicions that plenty more dirt lay in other unseen corners.

I thought Dose had guts, putting his principles in front of job security. Curious about the cutthroat trout in Squaw Creek, I shared my recollections of fishing there in 1965. Dose listened to me intently, and then said, glum-faced, "They're all gone now."

I didn't want to believe him. I couldn't believe him. I'd even entertained thoughts of going back to Squaw Creek to relive that wonderful experience. This bad news messed with my own personal dreamscape.

I sought some reassurance. Maybe he misunderstood which creek I meant; maybe I'd misunderstood him somehow. I pressed the point that this particular Squaw Creek emptied into Jackson Creek. He interjected, "I know the one. They're all gone."

I begged for some explanation. How could it be that in only thirty years the cutthroat population had gone from abundant to zero? These trout once lent testimony to a healthy, vibrant ecosystem. How could their loss be seen as anything but tragic? Dose said the cumulative effect of decades of heavy logging—including, no doubt, some timber sales that I'd worked on in 1965—along with recreational fishing in estuaries that pounded sea-run adults and the other usual culprit, fish hatchery practices—all this had rubbed out the natives.

Logging and other factors combined to eliminate a magnificent natural resource within my short lifetime. How could this desecration possibly happen?

Echoes of Leopold

In *A Sand County Almanac,* Aldo Leopold spoke to the larger issue that Squaw Creek represents: "...do we not already sing our love for and obligation to the land of the free and the home of the brave? Yes, but just what and whom do we love? Certainly not the soil, which we are sending helter-skelter downriver....Certainly not the animals, of which we have already extirpated many of the largest and most beautiful species....A land ethic...changes the role of Homo sapiens from conqueror of the land-community to plain member and citizen of it. It implies respect for his fellow-members, and also respect for the community as such."

Aldo Leopold paid attention to his world, exercising the simple virtues of observation and interpretation. He loved the earth and came to dislike and distrust the way humans treated it. As he grew disenchanted with the Forest Service, Leopold became increasingly marginalized within it. How else to explain his 1924 assignment to the Forest Products Laboratory in Madison, Wisconsin, except that he was cast aside from his land management responsibilities? Leopold remained at the lab until 1933, when he left the Forest Service to become a professor of game management at the University of Wisconsin, the first such professorship in the country. Leopold is credited with inventing the profession of wildlife management. He also helped found The Wilderness Society.

The Forest Service now proudly calls Leopold one of its own. I do not sense this was so in the 1920s, and even today, the values he espoused do not reside in the mainstream of Forest Service thinking and practice. How different might the Forest Service be today had Leopold become a prominent leader within the agency? I can't square what I read of his land ethic with the Forest Service we both worked for.

In 2005 the Forest Service produced a movie of the agency's first century.[18] I found the movie authentic; it genuinely, respectfully captured the essence of the agency of my experience. Orville Daniels, former Lolo National Forest supervisor, wryly noted in the video that the Forest Service "went over to the dark side" in its pursuit of timber. I squirm at the truth of his pithy reflection. I went there, too, for a while.

Daniels put it well, but what path took the Forest Service to the dark side?

The Forest Service's legacy comprises three main eras in its history. An initial custodial era typified the Forest Service from its creation in 1905 until about 1950. Then a utilitarian era emerged in support of strong demand for housing products following the Second World War. That period lasted until about 1990, when Judge Dwyer's spotted-owl ruling shut down old-growth harvest on national forest lands. The current era of ecosystem management, as yet ill-defined, remains a portent of things to come.

Historical evidence strongly suggests that once the Forest Service left the custodial era and shifted gears to embrace the priority of aggressive timber harvesting, it never looked back.

Who were the people making this shift? Most were men, veterans of the Second World War, the "greatest generation," lauded by Tom Brokaw, who approached the challenge of taming the forest with amazing determination and zeal. They mobilized for action and "got the cut out." Their passions united them with their predecessors—they loved the national forests and loved their work.

The utilitarians provided a muscular response from the national forests for a nation in need. The Forest Service believed fully in its ability to put these public forests to work for the good of a burgeoning country. America needed wood, and the Forest Service sat atop an enormous pile of it.

The Forest Service "outfit" built roads to access far-flung timber stands. The agency crafted contracts to enable the sale of federal timber to a rapidly growing sector of a timber industry that had no land of its own. This industry needed, and received, an assurance that, if it built sawmills, the Forest Service would supply the wood. Forest Service nurseries provided millions of seedlings for planting logged areas. The outfit fought fires so valuable timber didn't burn. The Forest Service stood tall...for a while.

Aldo Leopold noted the first rule of intelligent tinkering: keep all the parts. I cite Squaw Creek as a simple example of unintelligent tinkering. Larger examples include the controversy surrounding clear-cutting on the Monongahela National Forest and the flawed timber practices in Montana targeted in the Bolle Report. I doubt the foresters at Tiller who laid out the clear-cuts or the engineers who surveyed the roads anticipated anything bad happening to Squaw Creek. Today, we have come to a more organic way of seeing nature and its intricate relationships, but the prevailing dogma in 1960 was an industrial mindset. Foresters viewed forests as a machine that produced wood, and the Forest Service mission was to get the cut out. We simply did not discuss keeping all the parts.

But, while understandable, I don't think that's a reasonable excuse. I believe a good steward of the land must listen and learn. What does the land say? Nothing of what I've seen in Pacific Northwest landscapes would lead me to conclude that the best way to steward the land is to clear-cut the old-growth forest and punch roads into every nook and cranny. Unless, that is, you're "just logging."

History is replete with examples of economic forces run amok to maximize consumption or harvest. We humans have tended to take as much as we can, often far more than we should, regarding the resulting financial wealth as a reasonable reward for our exploitation. Think of the depletion of the bison, passenger pigeon, Atlantic salmon, bluefin tuna, sperm whale—the list is tragic and growing.

In my forestry analogy, wood consumption adhered to the same pattern. The logging industry worked its way across America, exhausting native forests along the way until it encountered the mother lode in the Pacific Northwest, home to some of the most magnificent forests in the world. There, logging of old-growth Douglas fir turned huge profits because the trees were enormous and the wood an excellent building material—workable, light, and strong. A single tree contained enough wood to build several homes. Early loggers simply sought the biggest and best trees to pluck out of the forest. The supply seemed inexhaustible. Inevitably, efficiency led to the practice of clear-cutting, and every tree fell to the saw. Clear-cutting was cheap and highly profitable. When small seedlings replaced the former giants, a plantation emerged where a forest once stood. But the plantation mimicked in mere caricature the complex, diverse, healthy ecosystem it replaced.

Thus did old-growth forests slowly diminish, and with them the spotted owls and salmon and hundreds of other species of plant and animal we've never heard of. I believe the Forest Service and timber industry were out of their depth, driving logging practices far beyond knowledge or any semblance of humility.

This mentality leads a chief like Dale Robertson to decry clear-cutting, yet claim in the same breath that "the Cascades are different." It leads agency leaders to do as little as possible to protect owls from logging or to clear-cut at timberline in Wyoming knowing we couldn't get seedlings to grow.

This was the dark side that Orville Daniels spoke of. I picture the Forest Service sifting through its mailbag and deciding that certain envelopes need not be opened. Those letters just nag us with disapproval of what we're doing. We've heard all the arguments before, but we've got a job to do, and we know how to do it. The Forest Service refused to conscientiously wrestle with this profound truth: that a significant segment of the public we'd sworn to serve had rejected our management of their public lands, and the land itself was telling us of its distress. The Forest Service tweaked a few things, but the fundamental approach remained unchanged.

Forest Service people had some right to be proud of how much we'd accomplished. I listened often as our great history and legacy were recounted. I even piled it on when I could. Yet I found little appetite for an authentic accounting of the negative side of the timber-first equation.

True, the Forest Service had its internal skeptics and critics, yet these tended to propound arguments emerging in professional journals, academic research, and activist opinions. Leaders in the Forest Service aggressively defended our timber mission against these rear-guard skirmishes from within. To an agency so strongly committed to timber harvesting, it seemed reasonable initially to dismiss contrarian views.

It took the environmental community to play the key role in revealing our dark side to the world. Opposition grew and mobilized as year upon year revealed more clear-cuts and more roads. Negative sentiments broadened, deepened, and persisted as many environmental organizations grew rapidly in size and influence and started hiring lawyers to defend their interests. The real war engaged the courts, using ambitious new laws like the National Forest Management Act, National Environmental Policy Act, Endangered Species Act,[19] and Clean Water Act[20] as weapons. The Forest Service was regularly beaten in court. A forest supervisor can shush a fisheries biologist, but not a federal judge.

When a federal agency stewards a resource in trust for a public constituency, a social contract exists. When trust in the Forest Service eroded sharply, we should have recognized that we had violated our social contract, that we needed to retool, that the most important thing for us now was to regain credibility and trust. Instead, the Forest Service battled for the status quo.

Judge Dwyer's 1991 ruling stopped federal logging of spotted-owl habitat and symbolized the death of industrial forestry on national forests. Dwyer's finding that government officials had willfully violated the law served as an exclamation point. The dark side was in the open now.

Finding Our Land Ethic

In the realm of human interaction with the environment, we are clamoring for sustainable conservation models that maintain forests while supplying vital human needs. Herein lays a tremendous opportunity— and challenge—for the Forest Service. The agency has a sizable and diverse land resource to manage, an ecologically oriented mission, excellent research capacity for technology transfer, and a workforce of thousands of the world's best and most dedicated professionals. In my view, the Forest Service is providentially positioned to provide an effective model of ecosystem management—it is the essence of the agency's mission.

My opinion, shared by many in the Forest Service itself, is that the agency is struggling to clarify its mission, mobilize for action, and make effective progress. Surveys identify poor leadership, stale morale, vacillation, chronic inefficiencies, and lack of competence as causes of widespread torpor. Many people I knew were more concerned with hugging the past than releasing it to fly into a more promising future. I give the Forest Service credit for improving some of its methods of doing business, such as collaborative processes that give interested parties a more direct role in decisions. That improves "how," but it does not directly address the larger issue of "what."

The key challenge is that the Forest Service has not yet embraced a land ethic capable of securing a more desirable future. It's past time to drive a stake in the ground and declare a future the agency aspires to. I believe in the value of taking a stand. I think my saying, "We will no longer clear-cut mature timber on the Siuslaw" enabled real change to occur. Like the wayfarer in Robert Frost's poem, we stand at a crossroads—knowing the history of the well-traveled road behind but needing to choose a less-traveled road to a better future.

The famous anthropologist Margaret Mead said, "Never doubt that a small group of thoughtful, committed citizens can change the world; indeed, it's the only thing that ever has." I take encouragement from this. I believe there are a few people with the acumen and the science and the management wisdom to move the Forest Service forward. Occupying a leadership position does not automatically make one such a person. But Forest Service leaders need to aggressively seek out the people with these answers, then mobilize for transformational change to secure sustainable forestry practices.

The transition of our public forests to timber production after the Second World War was a policy choice, enabled and led by the Forest Service. We can make a similar policy pivot to manage public lands primarily for diverse habitats, clean water, restoration, carbon stores, and other environmental values, while still producing wood products sustainably. As simply as I can put it, this means managing toward a more natural forest.

Timber harvesting need not cease. The Menominee Tribe in northeastern Wisconsin has been actively and sustainably managing timber on its 235,000-acre reservation for many generations. Tribal foresters do not clear-cut, and they do not maximize profit. Selective logging is designed to maintain or enhance environmental values. Wood products flow continuously from a mature forest environment that meets the tribe's ecological and economic goals simultaneously. I believe the Menominees have chosen a wise, humble path that is broadly applicable.

And consider again the Siuslaw National Forest. It is managed today under very different principles than those that prevailed twenty-five years ago. A critical tool we developed for that transition was stewardship contracting, designed to employ partners to achieve desired outcomes on a piece of land—reduce fire risk, improve wildlife habitat diversity, restore watershed health, and remove unneeded roads. Stewardship contracting is not logging for timber production. Rather, it is using the best means to manage lands for ecological sustainability. A few other national forests in the Pacific Northwest are beginning to emulate the Siuslaw.

Dramatic turnarounds are possible, even desirable occasionally. The Forest Service had its chance to turn around thoughtfully and deliberately, but it squandered this prerogative when it failed to respond to public cries for action. Now it is time to act quickly and boldly. We are not experiencing a blip or fad. Change is here to stay.

The Good Life

Jack Lessinger, a University of Washington economist, uses the term penturbia (denoting the fifth discernible wave of social settlement patterns in America) to describe areas beyond traditional urban and suburban zones. These, he says, are emerging as the best places to live and invest for people seeking the good life. I first met Lessinger in about 1995 when he spoke at a Forest Service meeting. He described his analysis of social trends and his finding that "urban escapees," as he called them, have a penchant for moving to nice forested places in pursuit of the good life. For the Forest Service, this is a highly pertinent development, because these folks become instant neighbors, with high expectations, when they settle close to national forests.

Furthermore, penturbia is associated with what he describes in the current era as "caring capitalism," which, he says, includes caring about the environment. Money, as always, continues to drive people's choices. But Lessinger argues that people increasingly see money that derives from assaulting or neglecting the environment as a bit dirty. Living in the woods beyond the cities symbolizes appreciating and valuing the environment.

As a forest manager, I contemplate the implications of Lessinger's research and I see people seeking the good life in communities like Bend, Oregon; Durango, Colorado; Bozeman, Montana; St. George, Utah; and Coeur d'Alene, Idaho. These towns bristle with newcomers as housing prices spike. Forest Service firefighters are well acquainted with the explosion of folks living in the woods and snuggling up to federal land that they hope and assume will remain undeveloped and, of course, free of fire. When these houses burn down, there will be blame.

This phenomenon of living the good life in the woods has occurred all over the country. However you may regard the attitudes and expectations of these newcomers seeking their good life at the forest's edge, the principle at work here is that the forest's environmental value ascends while its commercial value for wood products diminishes. Many forest managers, both private and public, also recognize this shift. I see emerging in forestry a new ethos that strives to develop economic, social, and ecological strategies aimed at growing forests that function more naturally, producing environmental services like clean water and air, abundant fish and wildlife, and even inspiration for the soul.

The committee of scientists charged with reviewing the National Forest Management Act in the late 1990s conducted numerous meetings all over the United States. The panel captured a strong, common theme in its March 1999 report, Sustaining the People's Land. Ecological sustainability, the scientists stated, should be the guiding star for our national forests. Economic and social elements are important, but these should remain subordinate to environmental integrity.

This trend has been increasingly evident in both public sentiment and the arc of numerous environmental laws governing federal lands in the latter half of the twentieth century. Society is clearly moving toward ecological values and away from narrow-purpose economic interests. These laws reflect society's heightened sensitivity to the environment and consequent expectations to improve the management of our shared natural resources. I disagree with those who feel the current suite of laws governing national forests is a jumble of conflicting, outdated mandates. A unifying thread of these laws is a role for the Forest Service as an advocate for environmental values.

In the Pacific Northwest and almost everywhere else, forestry practices have traditionally adhered to capitalistic economic principles. The environmental movement, in all its facets, speaks to this simple truth: our natural world does not, and cannot, operate solely on capitalistic economic principles, because in practice these principles are antithetical to environmental health. There's no future in taking every last stick of wood. Nature suggests that moderation matters when it comes to harvesting natural resources.

I am not arguing against consumption per se. From any rudimentary observation of the natural world, it is plain that consumption is normal and essential. I watch in fascination as humpback whales in Alaska work in concert to herd a shoal of herring for the payoff, when each whale devours thousands of herring in a single gulp. Even though the whales' consumption of herring seems almost wanton, the abundance of herring is stupefying, and many thousands escape.

Such abundance as this evokes for me the religious principle of the tithe—in giving, you receive. If consumption occurs at rates of, say, 10 or 30 or even 50 percent, the unconsumed remains assure continued abundant production (depending, of course, on a species' reproductive capacity). With moderate rates of consumption, this production-consumption continuum sustains itself even through the occasional stresses that temporarily diminish production.

What I'm objecting to is the notion of maximum consumption. Clear-cutting old-growth forests provided the greatest volume and value and returned the highest profit. But the maximizing of consumption put too great a stress on natural systems, and they crashed. Natural forests became tree farms, salmon populations plummeted because these forests were their womb, and all manner of old growth–dependent wildlife suffered.

Because nature is more complex than we can understand, we will no doubt discover other truths. For example, we are still accumulating research that points to the important role of mature forests in sequestering carbon. While I was on the Siuslaw, and again when I was developing the NFMA planning regulations for 2000, I don't recall a single mention of the nexus between forests, carbon, and climate change. Now we know that, globally, forests contain about one-third of the earth's carbon budget. Furthermore, the coastal temperate rainforests of the Pacific Northwest contain massive quantities of carbon, many times more per acre than the extensive boreal forests of Canada and Russia. We also know that old-growth forests contain far more carbon in their trees and soil than do young plantations. In the space of less than twenty years, national forests have gone from not even being discussed as valuable carbon stores to having a potentially significant role as a shock absorber for climate change.

Optimize, Don't Maximize

We need not choose between taking it all and leaving it all. Rampant exploitation for transitory profit operates at the extreme economic end of the spectrum, while preservation and exclusion of humans with no utilization of resources operates at the ecological extreme. A middle ground does exist where fundamental stewardship humbly, modestly, and sustainably utilizes what is produced and provided by nature. These beneficial and essential environmental services can be enjoyed in perpetuity as long we do not irreparably damage our forests, soils, water, and air.

In sum, instead of striving for maximization, we need to develop the more moderate and habitual practice of optimization. To achieve optimal outcomes, we must account for environmental costs that will necessarily result in less revenue and profit. Our baser instincts may lead us to want all we can get now—yes, instant gratification. Yet any reasonable person ought to prefer having enough forever. Wouldn't you rather achieve less profit perpetually than maximize short-term profits now and reap the grim result of bankrupt ecosystems later?

When I worked on the Siuslaw National Forest, we began to achieve optimization as we shifted away from the maximization model that had characterized the timber program from 1950 to 1990. The Siuslaw's management has evolved even further since then, to the point where I maintain that it is both ecologically and economically sustainable.

And in the Black Hills, where in 1975 I first saw the battle for the future engaged as Jim Hagemeier went toe-to-toe with prevailing dogma, some changes are apparent. On my last trip there in 2012, I saw evidence that recent forest management is following a more ecologically appropriate pattern. I saw relatively few large ponderosa pines sprinkled through a sea of waving grass, with not-too-crowded smaller pines populating the site as well. I looked on this as more reminiscent of the landscapes in Illingworth's 1874 photos. This was nothing like the Black Hills forestry I had observed—and helped achieve—in the 1970s. The foresters there told me that their purpose now is to create a healthy, more natural forest that was intended to be more resistant to catastrophic fire and bug infestation. In that regard, I think they have succeeded. Yet, I still detected too-strong passions about producing wood, and an insufficient commitment to growing and protecting an even larger number of big pines that would serve to anchor ecosystem health.

True change can be slow in arriving, but there is no mistaking it when it appears. It seems that Black Hills foresters are beginning to work more harmoniously with the ecosystem, rather than simply trying to impose a timber-production regime.

Their fundamental ethic seems to be changing, because they are now managing forests, not just trees. They are asking themselves: How can we design forest management so that it will enhance diversity and sustain a more natural forest, not just a tree farm? Can we manage so as to encourage frequent, beneficial fire, instead of creating a forest that can't tolerate fire? Can we design roads and bridges as if we actually expect the next big flood, instead of being shocked when it happens?

The Menominee Reservation and the Siuslaw and Black Hills National Forests have one thing in common: on-the-ground management that reflects intentional leadership. Leaders create a vision and then exercise prerogatives to bring that vision into being. Like any bureaucratic organization, the Forest Service tends to adopt normative behaviors and then resist change. Yet the organization is also large enough to permit and promote constant exploration and innovation. At its best, the Forest Service has an extraordinary agency culture that is capable of extraordinary things. As a result, forests like the Black Hills and Siuslaw have turned a corner and now travel a different road.

Let's recall that the conservation movement and the Forest Service itself were born from intentional, principled leadership. President Theodore Roosevelt and Gifford Pinchot, the first Forest Service chief, saw environmental abuse and did something about it, even though their actions were unpopular at the time. Both men crawled around the floor of the White House marking future national forests on maps. Then, by executive order, Roosevelt created the bulk of our existing system of national forests to beat a midnight deadline imposed by Congress to rescind his authority. The vision of Pinchot and Roosevelt on hands and knees seems almost comical, but it demonstrates their serious intentions.

The problem now is that the Forest Service's intentions for the next century and beyond are unclear. They need to be made clear if the agency is to excel. The first thing Forest Service leaders need to say— loudly, to internal and external audiences—is that both public senti- ment and science have shown that former management principles for our national forests are no longer valid. We tried the "timber is king" approach, and it failed. Restoration will be necessary and will be an ongoing priority. This can be done without repudiating former leaders and their motives.

Next, the agency's leaders need to explicitly embrace the mandate of ecosystem management, which I would describe as value-driven resource management with a goal of maintaining or achieving naturalness. Primary values should be clean water and air, abundant fish and wildlife, quality recreation opportunities, and sustaining landscape function.

The broad principles of ecosystem management can be applied to any landscape—longleaf pine savannah, northern hardwoods, mountain ponderosa pine, or West Coast Douglas fir. The specific methods will necessarily vary from place to place, but the intention should be to affirm the primacy of ecological sustainability for our national forests. If this concept of ecosystem management is still murky for some Forest Service people, leaders need to clarify its meaning with simple, understandable principles. For example: *We will work with nature, not against it. We will resist extremes and embrace moderation.* Jack Ward Thomas, Forest Service chief from 1993 to 1996, once said, "If you want golden eggs, don't kill the Golden Goose!"

Next, the agency's leaders need to create forest plans and take actions that are consistent with the ecosystem-management mandate. They need to make it clear that, while commercial activity is essential, maintaining environmental integrity is paramount. And they need to proceed with a sense of urgency.

Leaders of the Forest Service have the latitude and the authority to create change; that's what society expects of leaders. I say to them, "Don't squander the privilege and responsibility of leadership. Make a choice. Embrace it enthusiastically. Commit fully to creating a new reality and paradigm." If they do these things, the benefits will redound globally to other forests and those who manage them.

Many times in the past, in many circumstances, people like Roosevelt, Pinchot, Leopold, and Mike Dombeck have stepped up to a challenge. I have done it myself in my own small sphere. I reached a point in my career where I could go no further. I had changed, not in a moment, but in a season spanning my years in Washington, DC, and my early years in Oregon. I'd become increasingly disenchanted with a Forest Service—*my* Forest Service—that seemed far too beholden to industry interests. My early heroes met challenges head-on, but they got stuck in a rut when they failed to recognize that society's values had shifted, that the tools in their box no longer worked.

I met the challenge on the Siuslaw National Forest by building with new tools. The tools are good and the building continues and stands firm. This is a forest that is beginning to grow back into its natural self again. My greatest satisfaction in my long career is the feeling of taking something broken and putting it right.

Some might point out that I paid a price. Well, yes, I did. But that often comes with the territory. Pinchot was fired, Leopold neutralized, Dombeck run off. Yet the cost of advocacy is trivial when compared to what endures.

The finest tradition of American excellence calls for us to initiate change when conditions warrant and then to take bold action. The world waits for leaders to step out, without apology, to create a different future; create hope again, and again, and again. To the Forest Service I served and loved and had to leave, I implore you:

Do it now.

Acknowledgments

To lovers of national forests who I had occasion to meet throughout my career, thank you for sharing your values. You give life to the dreams of Teddy Roosevelt and Gifford Pinchot.

I had the privilege of working with and for dedicated, hard-working professionals who instilled a belief in the art of the possible. Thank you for firing my imagination.

I thank my brother-in-law Bill Ford, who coaxed this memoir out of me with encouragement and optimism in the belief that stories need to be told to be heard. You said the telling would be worth the effort. It was.

Mary Elizabeth Braun, acquisitions editor for Oregon State University Press, somehow turned a forester into an author, which is no small accomplishment.

Last, there is a reason authors acknowledge the gift of editors—things turn out so much better with their help. Gail Wells, with gentle ferocity, kept asking for more and better.

I begged her to just write the elusive parts herself; she knew what I wanted to say. She always refused.

Notes

1   Brokaw, Tom. 1998. *The Greatest Generation.* New York: Random House.

2   Kaufman, Herbert. 1960. *The Forest Ranger.* Baltimore: Johns Hopkins University Press.

3   Leopold, Aldo. 1949. *A Sand County Almanac.* Oxford: Oxford University Press; second edition (December 31, 1968).

4   Progulske, D. R. 1974. *Yellow Ore, Yellow Hair, Yellow Pine.* South Dakota State University Bulletin 616; first edition (1974).

5   Proulx, Annie. 1999. *Close Range: Wyoming Stories.* New York: Scribner.

6   Doig, Ivan. 1979. *This House of Sky: Landscapes of a Western Mind.* San Diego: Harcourt.

7   National Forest Organic Act. June 4, 1897. (30 Stat. 34–36; codified USC. vol. 16, sec. 551)

8   National Forest Management Act. October 22, 1976. (P.L. 94–588)

9   Multiple-Use Sustained-Yield Act. June 12, 1960. (P.L. 86–517)

10 Dwyer, W. L. 1991. Order on motions for summary judgment and for dismissal. Seattle Audubon Society et al., plaintiffs, v. F. Dale Robertson et al., defendants. No. 89099 (t)WD. Seattle, WA: US District Court, Western District of Washington.

11 Clinton, W. J. April 2, 1993. "Remarks on Opening the Forest Conference in Portland, Oregon."

12 National Environmental Policy Act. January 1, 1970. (P.L. 91–190)

13 USDA, Forest Service. 1972. Roadless Area Review and Evaluation

14 USDA, Forest Service. 1979. Roadless Area Review and Evaluation II

15 USDA, Forest Service. 2001. H. Gucinski and M. Furniss, eds. "Forest Roads: A Synthesis of Scientific Information."

16 Johnson, K. N., et al. 1999. "Sustaining the People's Land." (Federal Register Vol. 65, No. 218)

17 Maclean, John. 2007. *The Thirtymile Fire: A Chronicle of Bravery and Betrayal.* New York: Henry Holt and Co.

18 USDA, Forest Service. 2005. The Greatest Good: A Forest Service Centennial Film.

19 Endangered Species Act. December 28, 1973. (P.L. 93–205)

20 Clean Water Act. October 18, 1972. (P.L. 98–500)

Index